"From morality to molecules, environment to equity, climate change to cancer, and politics to pathology, this is a wonderful tour of global health – consistently presented in a clear, readable format. Really, an important contribution."
—Professor Sir Michael Marmot, *Director, Institute of Health Equity, University College London*, Author of "The Health Gap"

"Paolo Vineis provides convincing evidence that our failing economic systems and the plundering of our environmental resources are threats to health that cannot be combatted by leading a healthy life-style or by the efforts of national governments. A 'must read' for politicians, skeptics and those of us who are attempting to promote international collaboration in environmental protection, economic strategies and public health.

This book has a stark message—if we carry on as usual, the global gains in health may be reversed. Read how we got into this mess and how we may be able to get out of it by means of international economic and environmental collaborations. Vineis' book is a major contribution to global health thinking which should influence policy and practice."
—Professor Shah Ebrahim, *London School of Hygiene & Tropical Medicine*

"This is an excellent and timely introduction to the most important health issues of our time. Global health is presented as a 'well-conceived thriller' with many actors and an uncertain ending. This is an excellent overview of the rapid global changes we are facing in our environment, their consequences for health, and the potential solutions. Should be essential reading for all those involved in global health, health research and health services, as well as all those concerned about the health of the planet and its inhabitants."
—Neil Pearce, *Head of Department of Medical Statistics and Professor of Epidemiology and Biostatistics, London School of Hygiene & Tropical Medicine*

"This book is a salutary and soundly argued reminder that the 'common good' is not simply what remains after individuals and groups have appropriated the majority of societal resources: it is in fact the foundation on which any society rests and without which it collapses.

Strengthening this foundation is vital to successfully tackling the growing global environmental and economical challenges, which are unequally distributed between and within countries. The book emphasizes that for this purpose technical solutions based on political choices inspired by social solidarity are called for. They represent feasible alternatives to the dismal and hazardous prospects that the survival of human-kind will lie in changing humans into trans-humans through biotechnological manip-ulation, or in the escape to Mars or an exoplanet for a very few people out of billions."
—Rodolfo Saracci, *Senior Visiting Scientist, International Agency for Research on Cancer, Lyon*

Paolo Vineis

Health Without Borders

Borders

Epidemics in the Era of Globalization

Copernicus Books is a brand of Springer

Paolo Vineis
Imperial College
London, United Kingdom

ISBN 978-3-319-52445-0   ISBN 978-3-319-52446-7   (eBook)
DOI 10.1007/978-3-319-52446-7

Library of Congress Control Number: 2017941007

Printed on acid-free paper

This Springer imprint is published by Springer Nature
The registered company is Springer International Publishing AG
The registered company address is: Gewerbestrasse 11, 6330 Cham, Switzerland

Europe has every right and every capacity to be able to offer the finest model of social welfare on earth: we must stop squandering our opportunities!
Thomas Piketty
New York Review of Books, February 25, 2016

This book is dedicated to Ludwik Fleck, a great philosopher and scientist, and Tony McMichael, a great epidemiologist and a master of integrity, in the hope that future generations will read their work.

Preface

Why This Book

Like in a well-conceived thriller, many different trends in the world are now converging to create an almost unbearable suspense about what our future holds—although the plot remains blurred, it is clear that life has become increasingly unsustainable and uncertain for many. Climate changes, mass migrations, wars, the economic crisis, pandemics, the epidemics of obesity and diabetes and unprecedented environmental degradation can no longer be considered separate threats to be dealt with in isolation. Rather, these and other global threats to human health and well-being often share common causes and have impacts that are converging dangerously towards an unknown future. The consequences for human health have not been fully described and remain poorly understood. The principal motivation behind this book is to attempt to draw some of these threads together into a more coherent picture, to understand what the evidence suggests about the emergence of multiple health threats from global changes and their interactions. The guiding hypothesis is that we have all contracted a **double debt: one is financial (and is well known) and one is environmental (and is underestimated, particularly in health circles, and more serious).** A second theme that is explored throughout the book is that, like the economy, the state of health of humankind—that on average has always improved so far—may become increasingly unstable with areas of decline, accompanied by an ongoing widening of the socio-economic divide particularly within nations.

After an introductory chapter that explains what "global health" is and describes the intertwining of the threats that the planet is facing, the following chapters are devoted to the main topics that contribute to global health; that is, the impact of globalisation on the onset and distribution of diseases. These chapters address, for example, food, climate, environmental pollution and the impact of the economic crisis on health. A separate chapter is devoted to cancer because its many and varied forms epitomise the multidimensional aspects of the problem: the divergence between increasing incidence and decreasing mortality rates in rich but not in poor countries (where both trends are increasing); the role of improved diagnosis; differential access to therapies; the changing landscape of risk factors; and the political problems of cancer control. Another chapter addresses a fundamental recent discovery about the mechanisms of disease, which promises to be of great relevance for global health: "epigenetics" provides a new perspective on the ability of the environment to modulate the expression (functioning) of genes. In an era of constantly changing environmental landscapes—like in the quality and composition of food—epigenetics may prove to be essential in mediating the health effects of environmental change. The last two chapters draw some conclusions on how policies could be directed towards creating potential solutions for the problems described, and how only redefining the environment and health as "common goods" requiring our concerted protection can be effective in sustaining quality of life in the medium to long term. It is obvious to many that waiting for potential solutions to emerge from the laws of the market alone is not going to be enough to meet the health challenges of the twenty-first century. If the environment and health are not eventually conceived and appreciated as common goods of humanity, then the separate problems described in this book will not be resolved.

London, UK   Paolo Vineis
September 2016

Acknowledgements

I am particularly grateful to Shah Ebrahim and Elisabete Weiderpass for their very thorough reading of the manuscript and the many suggestions and stimuli they gave me. I am also very grateful to Rodolfo Saracci, David Blane, Kris Murray, Cristina Savio, Enrico Casadei, Roberto DeVogli, Pietro Ortoleva, Pietro Terna, Nerina Dirindin, Luca Carra and Roberto Satolli for their attentive reading and precious advice.

Part of this book is taken from the lectures on the course of Global Health that I have coordinated for the past seven years (together with Helen Ward, Graham Cooke and Mariam Sbaiti) for the students of Medicine at Imperial College of Science, Technology and Medicine, London. The following link gives access to part of the course contents: http://tinyurl.com/pnlagc7.

I thank Demran Ali and Shahiba Begum for their work on Nauru at the BSc in Global Health at Imperial College in the years 2009–2011.

Contents

1

The Double Debt: Economic and Environmental

The Economic Crisis

It is premature to establish whether the recent economic downturn has had an impact on health, but some comparisons between countries that have adopted different policies may be useful. It is probably still impossible to disentangle the effect of the crisis itself from the impact of the measures that have been put in place to cope with it. Obvious consequences of the crisis are a reduction in private and public expenditure for health including preventive activities such as screenings or the increase in sales of poor quality and cheap ("junk") food that increases the risk of obesity and diabetes. The direction of changes is not always negative, however: for example, the consumption of alcohol decreases with an economic downturn (Hessel et al. 2014; see also below, Chap. 2). In the UK, the combined effect of the crisis and political choices has generated new forms of impoverishment, as documented in the book "Breadline Britain" (Lansley and Mack 2015). Just to give a few figures, in 1983 14% of families lacked three or more essential necessities (according to the standard classification used to define the poverty threshold), whilst this rose to 30% in 2012, and between 2002 and 2013 food prices rose by over 50% whilst inflation rose by about one-third. The UK is very similar to the United States in being a socially unstable society, where the alleged low unemployment rate is in fact based on a large proportion of low-paid workers (25% in the USA, 20% in the UK, as opposed e.g. to between 5% and 10% in Finland or Norway). The impact on health of contemporary increases in inequalities and of the shrinking role of the Welfare State in the UK is yet to be studied, whilst health inequalities in

© Springer International Publishing AG 2017
P. Vineis, *Health Without Borders*, DOI 10.1007/978-3-319-52446-7_1

the UK have been largely investigated in the past, starting at least in the 1980s with the work of Douglas Black and then Michael Marmot. The recent economic crisis seems to be so deep that there has been a proliferation of "food banks" in the last few years, i.e. a large number of families live on food donations. And of course the food that is eaten by poor families is not the healthiest. The UK and the United States, two countries where I have lived (in the UK for more than 12 years), are a patent example of food inequality, with a widespread consumption of cheap junk food in the lower social strata.

The Environmental Debt

The debt towards Nature (due to the loans "in kind", like extensive extractions and overconsumption of natural resources) is likely to be much larger than the economic one, and more difficult to pay back and to audit, if not impossible. The use of land and the related erosion and impoverishment, the misuse of water, fisheries or the atmosphere, just to give a few examples, are creating a debt with Nature that will not be able to be paid back, even in decades. The multiple aspects of the environmental crisis have just started emerging, as the report from the Lancet-Rockefeller Commission has made clear (Whitmee et al. 2015).

It is difficult (and increasingly so) to identify priorities for our environmental and social choices, i.e. the problems that need to be addressed first amongst the many that are looming. This book takes an "equity" angle, i.e. it looks first at the most disadvantaged in society, both within and between countries. By equity angle I mean that inequalities are unjust but also amenable to social/political interventions. From this point of view, the greatest challenges are likely to concern food and water, their accessibility and their price. A potentially devastating problem related to globalisation and exacerbated by climatic change is represented by fluctuations in food prices. One example is the Russian drought and heat wave of 2010, when 17% of the total crop area of the country was affected, causing a national drop of 33% in the wheat harvest. As a result, a ban on wheat exports was implemented that led to a 60–80% increase in global prices in the summer of 2010 (Grebmer et al. 2011). This resulted in major increases in wheat and bread prices in importing countries such as Egypt, Syria and Yemen. The problem is also related to the financial drift of the economy, i.e. the fact that staple goods such as food are subject to financial speculation. This—together with dependence on food imports—leads to fluctuations in prices and increases the vulnerability of low-income countries.

Concerning food, another aspect that emerges and is typical of globalisation is the liberalisation of trade on the basis of international treaties, whose meaning is still equivocal. The general framework for such treaties is provided by the WTO (World Trade Organization), whose aim is to protect free trade by reducing or abolishing barriers: tariffs, but also barriers deriving from laws that protect the health of workers and consumers. Officially, the WTO acknowledges the need to protect health, but in fact it has opposed measures that limited tobacco sales, interpreted as threatening the freedom of trade. This was the case, for example, in relation to the Framework Convention on Tobacco Control (as we show later in the book) and the WHO (World Health Organization) Global Strategy on Diet, Physical Activity and Health. The international treaties open up the possibility for multinational corporations to sue governments if their rules or laws limit the freedom of trade. This is an obvious coercion of the rights and autonomy of governments and their citizens and creates the paradoxical situation of a lack of multinational governance for climate change together with the increasing—and in fact global—legal power of large companies (Baker et al. 2014; Stuckler et al. 2012). This has already occurred through bilateral treaties: for example, Philip Morris has challenged the use of dissuasive messages on cigarette packages on the basis of a treaty between Uruguay and Switzerland. Similarly, the Australian policy of "blank" packages has been attacked on the basis of a treaty between Australia and Hong Kong. An analysis of the sales of certain goods in countries that have signed bilateral treaties with the United States shows that for example the consumption of sweetened beverages was 63% higher than in non-signatory countries (Baker et al. 2014). The NAFTA treaty between the United States and Mexico was accompanied by a significant increase in sales of sweetened beverages in the latter, a country that currently has the highest consumption in the world (300 L per capita per year) and one of the highest frequencies of obesity (Baker et al. 2014). There are many other examples, like the more tolerant attitude towards hormones in meat and endocrine disruptors in the United States compared to Europe. Even the national health systems could be threatened by the treaties, because governments could be sued for limiting opportunities for profit in medical care.

These views are not simply political; they are shared by a large part of the medical profession. For example, the journal of the Faculty of Public Health in the UK (Royal College of Physicians) has recently published an article with the telling title "Warning: TTIP could be hazardous to your health" (Weiss et al. 2015) (TTIP is the Transatlantic Trade and Investment Partnership).

The Baumol Disease and the Funding of the State

Another global problem concerns the booming costs of health care. As the Director of the International Agency for Research on Cancer (www.iarc.fr), Christopher Wild, has repeatedly written, we cannot "treat our way out of cancer", i.e. the steep increase in the number of cancer cases in developing countries is simply incompatible with the very high costs of therapies. If we consider the United States alone, health care is by far the largest industry (including insurances), employing more than 10% of the workforce, and it is one of the largest causes of expense for the average family. Clearly, this cannot be afforded, particularly if one thinks of the poor performance of the American health system in terms of outcomes and inequalities. Obamacare is a partial and limited response to the healthcare crisis. Unfortunately, as for many other issues, Obamacare has been attacked by an aggressive right wing which is often supported by the very same victims of the current system (a paradox that is related to the way the media address the problem, but that is another story).

One of the common criticisms is that public health care absorbs a lot of money and is improductive. Is this really true? According to a famous example of why the notion of productivity may be misleading, it does not make much sense to increase the productivity of a string quartet, e.g. by substituting musicians with machines or increasing output in other ways (Atkinson 2015). Humans cannot be always replaced, and this has far-reaching implications, in particular for the costs of the State services, i.e. for example education and health care. "Baumol's cost disease" is a phenomenon described by the economist William Baumol and defines a relative increase of salaries in jobs that experience no increase of labour productivity. Productivity grows faster in certain sectors, usually because of replacement of humans with machines. Salaries tend to grow in all sectors; however, some cannot enjoy growth in productivity because they cannot easily replace humans with machines or increase output in other ways. Costs and wages of the public sector, like in education, health care and disease prevention, thus necessarily increase in relative terms compared to more productive sectors. According to the example above, the same number of musicians is needed to play a Beethoven string quartet today as was needed in the nineteenth century; that is, the productivity of classical music has not increased. However, wages of musicians have increased greatly since the nineteenth century. This is at the basis of the fiscal crisis of the State and does not seem to be understood by more and more frequent opposers of the State.

Another important reason for underfinancing of the State activities is tax evasion. About 8% of the world's wealth (>7.6 trillion dollars) is held in tax havens, and, even more stunningly, this money proportionally comes more from developing countries than developed ones (Sunstein 2016). Russia—as in many fields including health statistics—scores very badly, with as much as 52% of its wealth sent abroad (Sunstein 2016).

One of the main theses in this book is that we need to seriously reconsider the role of the State and—in fact—of international organisations if we want to avoid health catastrophes in the next decades. An obvious implication is the revision of the currently ineffective world's fiscal policies. Side advantages of a transparent fiscal policy and the ban of fiscal havens would be combating the financing of terrorism and money laundering.

Co-benefits of Climate Change Mitigation

I alluded to the difficulty of setting priorities in a constantly changing scenario. However, not only forward-looking policies can be much more effective than short-sighted policies, but systemic approaches can be much more effective (and cost-effective) than sectorial interventions. One example is the interplay between climate change mitigation and health benefits. It is sometimes claimed that addressing climate change with proper policies is too expensive and can lead to a further decline in the economy. However, co-benefits of implementing climate change mitigation strategies for the health sector are usually overlooked. The synergy between policies for climate change mitigation in sectors such as energy use (e.g. for heating), agriculture, food production and transportation may have overall benefits that are much greater than the sum of single interventions (Haines et al. 2009). Here, I describe a few examples of climate change mitigation strategies that have important co-benefits for global health.

1. The transportation sector is often the single largest source of greenhouse gas emissions in urban areas. Policymakers have tried to reduce this emission by discouraging car travel and promoting other means of (active) transport. Active transport, such as cycling and walking, increases daily physical activity, and physical inactivity is one of the leading causes of non-communicable diseases all over the world. It has been estimated that the combination of active travel and lower-emission motor vehicles would produce great benefits for health, notably from a reduction in the number of years of life lost from ischaemic heart disease (up to 10–19% in London

and 11–25% in Delhi according to Woodcock et al. (2009)). Obesity, which is increasing dramatically all over the world, particularly in children, could also effectively be reduced by a more active lifestyle: a 30 min walk per day could—in many cases—be enough to even out small positive energy balances (though the relationships between overweight and physical activity and the direction of causality are still controversial).

2. Improving heating and cooking systems—for example by making them more efficient—reduces energy consumption. Improved models of stoves (electrical vs. biomass) allow a 15-fold reduction in the emission of particles and other pollutants in the atmosphere. Especially in developing countries—where old stoves are common—these improvements could also have a considerable positive impact on health: cooking on simple wood or coal stoves currently forms a major source of indoor pollution and increases the risk of certain chronic diseases, such as COPD (Chronic Obstructive Pulmonary Disease). Air pollution is the biggest environmental cause of death worldwide, with household air pollution accounting for about 3.5–4 million deaths every year (Gordon et al. 2014). The Indian stove programme installed hundreds of thousands of new stoves starting in the 1980s, and similar programmes have been launched in China, Indonesia and other countries with variable success (e.g. recently an investigation funded by the UK Medical Research Council found that the Malawi programme was not effective in reducing children pneumonia). This can lead to the prevention of a large number of deaths from COPD and other diseases if the programme is sustained in the long run.

3. Meat production is highly inefficient energetically: it requires an extremely high use of water and land per unit of meat. One-fifth of all greenhouse gases worldwide is related to methane production from livestock farms. Reduction of meat intake by consumers would lower meat production and is therefore often promoted as a climate change mitigation strategy. A high intake of meat is also associated with an increased risk of disease, in particular for certain cancers and cardiovascular disease.[1] Reduced meat consumption would therefore also have a major impact on public health. It has been estimated that a 30% reduction in livestock production in the UK would reduce cardiovascular deaths by 15% (Friel et al. 2009).

4. Non-renewable energy production, for example coal burning, is a major contributor to worldwide greenhouse gas emissions. Many countries have adopted policies to reduce polluting energy production and stimulate

---

[1] World Cancer Research Fund guidelines: http://www.greenhillosteopath.co.uk/documents/RecommendationsBooklet.pdf

production of (renewable) energy through cleaner sources. For example, since 2000, the government in the Chinese province of Shanxi has promoted several initiatives (including factory shutdowns) with the goal of reducing coal burning emissions. The annual average PM_{10} concentrations decreased in the large city of Taiyuan (Shanxi) from 196 µg/m (Lansley and Mack 2015) in 2001 to 89 $µg/m^3$ in 2010, which is actually still very high by Western standards. It has been estimated that the DALYs (Disability-Adjusted Life Years) lost in Shanxi had decreased by 57% as a consequence of the measures (Tang et al. 2014). The IPCC (Intergovernmental Panel on Climate Change) fifth assessment report stresses that the main health co-benefits from climate change mitigation policies come from replacing polluting sources of energy by renewable and cleaner sources, with a considerable effect on the improvement of air quality.

Final Remarks

The co-benefits from climate change mitigation for the health sector have not yet been completely identified and quantified, but they are an example of how intelligent and forward-looking policies could bring great benefits to the Planet and help reduce the double debt (economic and environmental). The topic of co-benefits does not appear on the priority list of political discourse: relevant sectors, including those involved in non-communicable disease prevention (Pearce et al. 2014), transportation, agriculture, food production and climate change (Alleyne et al. 2013), still work separately, whilst collaboration would enhance the synergy between health improvement and climate change mitigation and maximise benefits for both. It is hoped that these issues will be seriously considered in the follow-up of the COP21 Paris conference (that took place in December 2015), and co-benefits will become a common language on the political agenda.

One of the main theories of this book is that we cannot avoid taking a "common good" perspective when dealing with health (we clarify later what we mean by the expression "common good"): the crux of the problem is that an entirely profit-driven system is increasingly showing its distortions in all fields. Even worse, a competition-driven economy cannot—by its very nature—address global challenges that require strong international cooperation. Certain problems cannot be addressed in other ways except by assuming a global perspective rather than a national perspective. For example, it is unlikely that climate change mitigation could be seriously undertaken at a national level, because no country has an economic interest in doing so. After the Kyoto, Copenhagen, Durban, etc. summits on climate, the carbon–GDP ratio did not decrease significantly (with recent evidence, however, of slowing down of emissions in 2014).

Apart from climate change—but connected to this—there are many other initiatives that could be taken to protect and improve health, before it is too late. The Faculty of Public Health of the Royal College of Physicians of the UK has recently released a manifesto, the topics of which are extensively discussed in this book.[2] Their view that I fully share as a member of the Faculty is that—amongst other measures—a sugar tax, a minimum unit price for alcohol, restrictions on junk food marketing, a living wage and a zero carbon energy policy are the only ways to overcome the current health crisis and the increasing costs of health care. One extreme theory, discussed in a chapter of this book, is that life expectancy might decrease in the United States in the future because of the epidemic of obesity and diabetes, as claimed in a 2005 paper (but not supported by other evidence). A similar hypothesis is now reported from the European branch of the WHO: if on the one hand, Europe is faring very well and is reaching many of the health goals that had been set in the past, projections suggest that future generations may not attain the same life expectancy of their grandparents because of the joint action of tobacco, alcohol, diet and obesity (https://euro.sharefile.com/d-s6b26e8dff7042a9b). However, other recent projections suggest an opposite and more optimistic trend for the 35 most developed nations (Kontis et al. 2017). In a nutshell, long-term realism is needed to tackle global health (Bray et al. 2015).

---

[2] Start well, live better: a manifesto for the public's health. The Faculty of Public Health of the Royal College of Physicians. Available on www.fph.org.uk

2

An Overview of What Global Health Is

The state of health in the world varies enormously from one area to another and cannot be traced back to a simple diagram. On the one hand, there have been significant improvements over the decades: the average global life expectancy since 1960 has increased from just over 50 to over 71; smallpox has been wiped out and the number of deaths from measles, 871,000 in 1999, rapidly fell to an estimated figure of 12,000 in 2012: one of the undeniable successes of those vaccinations which today are erroneously opposed by sectors of the population. At the same time, the inequalities in the health field are also greater than in the past: there are more than 7 billion people in the world, but only 1 billion can expect to lead a long and healthy life. Whilst the life expectancy of a child born in Japan is 84, that of a child born in Swaziland is only 40; the maternal mortality rate is about 500 every 100,000 births in south-Saharan Africa and less than ten in many European countries.

Unlike in the past, health problems are no longer confined within the individual countries. The SARS epidemic, which started in China in 2002 and rapidly spread all over the world, recorded 8422 cases and 916 deaths in 29 different countries in 8 months. The emissions of CO_2 in the United States and China may be causing an increase in floods in Bangladesh. This means that control programmes, to be effective, have to be global, but it also means that the relations between health and the economic and political trends of globalisation are much more than superficial.

Global health, an expression that has now entered common use to refer to the specific problems linked to globalisation, has been defined by the United States Institute of Medicine as all those "health problems, issues, and concerns

© Springer International Publishing AG 2017
P. Vineis, *Health Without Borders*, DOI 10.1007/978-3-319-52446-7_2

that transcend national boundaries, may be influenced by circumstances or experiences in other countries, and are best addressed by co-operative actions and solutions."

The concept of global health is connected with another important phenomenon, that of the so-called epidemiological transition, or the continuous process according to which some diseases decline (many infectious diseases and those caused by malnutrition) and others spread (chronic "non-communicable" diseases). Although infectious diseases are still a major public health problem, non-communicable diseases are becoming the main cause of death also in low-income countries. Underlying these transformations there are biological, environmental, social, cultural and behavioural factors; furthermore, different stages of transition can coexist in the same community.

When the ranking of the main causes of death at global level is considered, the diseases that are more common in low-income countries are found in the fourth (lower respiratory infections), the sixth (HIV/AIDS) and the seventh (diarrhoeal diseases) positions (2012 WHO statistics). All the other causes of death—together with the risk factors traditionally referred to affluent societies (obesity, little physical exercise, high blood pressure, smoking and unbalanced diets)—have also become important in the rest of the world. Coronary diseases are the cause of 17% of deaths in high-income countries and of 11% of those in lower-middle income countries, and in both they ranked first as the cause of death in 2012.

According to many indicators, including health, there is a general convergence of the countries in the world (see http://ourworldindata.org/data/economic-development-work-standard-of-living/human-development-index/). Inequalities themselves tend to be greater and greater within countries rather than across countries (Bourguignon 2012).

Economy and Health

The influenza A (H1N1) epidemic in 2009 was one of the phenomena most clearly linked to globalisation: it started from pig farms first in the United States and then in Mexico, but soon took on global dimensions; defined a "pandemic", by early 2010 it had already caused at least 18,000 deaths. One interesting aspect for the theories in this book is that the H1N1 pandemic was also taken as a model by economists to describe the crisis of liquidity triggered off when the speculative bubble of the US real estate market burst. From then on, the epidemic spread of some infectious diseases, especially in the light of the mechanisms of contagion, has become a paradigm for economists wishing to describe financial crises.

Certain economic phenomena have been parallel to infectious epidemics, but the point is that both can be described with similar models, usually describing a localised crisis that rapidly spread all over the world. For example, as early as 1997, the devaluation of the Thai baht had given rise to a progressive destabilisation of the markets in South-East Asia, which then spread to Latin America, then Russia and finally to the United States, in exactly the same way as an epidemic. This crisis had coincided with the spread of another epidemic virus, the H5N1, known as the bird flu virus and which had also started in South-East Asia.

The most complete theorisation of the convergence of epidemiological and econometric models was made by Andy Haldane of the Bank of England in 2009, when he compared the outbreak of the SARS epidemic in Guandong (China) in 2002 with the bankruptcy of Lehman Brothers in 2008. Both episodes had some structural characteristics in common, which could be described with equations and which were the consequence of the "stress of a complex, adaptive network".

From these metaphors which brought together the economy and epidemiological models, there also came suggestions for repairing actions, in which the strategic regulation of the markets was compared to actions of public health. In particular, the theoretical physicist Robert May, well known for his work with the epidemiologist Roy Anderson of Imperial College, proposed an ecological approach to reform the banking system: in ecosystems characterised by complex interactions, the bonds that connect different species (e.g. men, birds and pigs) can trigger off moments of severe instability; similarly, mergers and perturbations in the financial system can spark off uncontrollable chain reactions.

Converging Trends: The Quest for a Minimal State, Misconceptions on Science, Fragmented Research

Together with the theories of the free market, a minimalist conception of the State has spread, implying that the State should interfere as little as possible with citizens' choices and decisions. When the (Republican) Mayor Bloomberg legislated on the portions of sugary drinks and prohibited trans-fatty acids in New York, criticism was poured on him by the defenders of individual freedom. To govern the rapid changes in technology and in the models of distribution and consumption, strong public institutions are necessary, but they have never been so discredited and unpopular as today. Today, strong international bodies are needed even more than national states: the agencies that represent public health ought to be more incisive than the WTO (World Trade Organization) when it is a question of legislating on products that are harmful for health like smoking. Research ought to be more incisive than the capacity of industry to confuse public opinion and orient its behaviours. Unfortunately, conflicts of interest have become a central component on the scene of relations between science, technology and politics.

Alongside the discredit of the public sector, we are also witnessing the spread of harmful anti-scientific prejudices, such as the opposition to vaccinations. The extreme cases of recrudescence of poliomyelitis in Nigeria and in Pakistan a few years ago—due to the prejudice according to which the Americans were alleged to be vaccinating Muslim populations to weaken them—showed how obscurantism and disinformation cause damage which can spread like wildfire. However, even in Western countries conservative movements such as the Tea Parties and other "alternative" movements are promoting anti-scientific and harmful opinions.

Another phenomenon that has to be taken into consideration is the fragmentation of scientific research. Human footprint on the Planet has now reached such an extent that we cannot afford research that is unstructured and motivated by pure scientific curiosity. In one of the chapters of this book, we will examine what the consequences of climate change can be (and to a great extent will be) on health: these consequences are too complex to be left to competition between researchers who study some aspects of them in a fragmentary fashion. Naturally, there are also some praiseworthy examples of cooperation and coordination on a very wide scale, in particular thanks to private foundations such as the Bill and Melinda Gates Foundation or the GAVI Alliance (Global Alliance for Vaccines and Immunisation). The latter is a very interesting model of public–private cooperation and is helping take vaccinations to children in poor countries all over the world (http://tinyurl.com/ljx5mpm).

The changes in the technologies at our disposal today and therefore in behaviour and lifestyles are often faster than the capacity of scientists to study their consequences. In the past 30 years, there has been a revolution in the food sector and the production of food has now reached a very high level of industrialisation, yet the epidemiology of chronic diseases still uses old tools and deals with single nutrients rather than the changed models of consumption. The new technologies of communication, from computers to smart phones, have spread more quickly than any other technology in the history of humanity, but research is still focused on a less central topic such as the presumed carcinogenic effect of electromagnetic fields whilst we have little concrete data on the cognitive and behavioural impact of these new technologies.

Even in the cases where research has investigated important topics, its translation into public health measures is opposed by powerful defensive reactions, especially from industry. For a cause of disease that we know almost everything about by now, cigarettes, global public health has to reckon with resistance by the WTO and its call for a free market, a topic that we will develop extensively.

In a constantly changing scenario, dictated by the increasing speed of trades, the role of the State and central institutions as a guarantee for citizens needs to be empowered.

Political Crisis, Health Crisis

The economic crisis has led to a lowering of the living standards of millions of people and it has been accompanied by an even deeper crisis of politics, with players who often have little credibility or are corrupt. There are signs of a deterioration in the state of health not only in the countries more directly hit by the crisis (Greece) but also in economically strong ones such as the United States. We cannot rule out that in the future many of the conquests in the health field may be eroded and that we will see a worsening of the state of health of large sectors of the population. The health crisis could be outlined through mechanisms which are not too dissimilar from those that led to the economic collapse: (1) the concentration of capitals in a small number of large corporations (in particular pharmaceutical); (2) a finance-based economy even in the health field, with the propensity to disinvest from the least profitable sectors; (3) sacrificing stable and political institutions that can have both a global and a local impact (as in the case of finance with Bretton Woods and with the guarantee systems that emerged from the Second World War, as explained later); and (4) the growing pressure on States to take on measures limiting expenditure and not to hinder investments and consumption, even though they may be harmful for health.

The victory against smallpox in the past was possible thanks to the strong collaboration between the World Health Organization (WHO) and the individual countries: today this pact risks fracture, to the detriment of other diseases.

Health in a Global World

One of the most visible effects of the three concomitant pressures of the free market ideology, of the economic crisis and of globalisation is the increase in the gap between social classes. It is also clear based on simple health indicators such as mortality. In general, men die earlier than women, but in both genders people with higher levels of education have a far longer life expectancy than the others (for Europe, see Gallo et al. 2012). In addition, in the past few years, the social gap in the most developed countries has widened for many of the health indicators even though it has narrowed for others (Bleich et al. 2012).

These simple observations raise important questions, including theoretical ones, which have not yet received clear answers. Does globalisation improve or worsen the state of health of populations? At the level of population, does health have elasticity, or in other words, does it resist the forces that threaten it, or not? And where is the "point of breakage"?

Health in the more developed countries improved extraordinarily between the end of the Second World War and the 1970s and 1980s. After that the *rate* of improvement slowed down (though life expectancy itself always increased in high-income countries), whilst rapid improvement has moved towards Brazil, Russia, India and China (grouped together under the acronym BRIC). It is likely that the extension of such improvement to Africa, in particular sub-Saharan, will follow with a certain delay. There is a clear parallelism between the time trend of economic indicators and that of the health macro-indicators, in particular of life expectancy. In other words, when the wealth of countries increases, the conditions of health in them also improve. What are the mechanisms underlying this relationship is still unclear, as a debate based on the seminal 1975 paper by Samuel Preston made clear (Bloom and Canning 2007). Preston examined the relationship between life expectancy and income in three different decades, the 1900s, 1930s and 1960s: in each decade the association between the two measures held true. Also more recent research shows that the income–life expectancy relationship still applies and continues to move upwards. One interesting aspect of the debate on Preston's paper concerns the possibility of a health-to-wealth causal relationship, i.e. not only wealth can lead to better health, but also the reverse may happen, through increased productivity: this could have great repercussions on our idea of health as a common good.

Other indicators, however, show less obvious trends and sometimes without clear explanations. We will deal mainly with these: the obesity and diabetes epidemic that is affecting almost the whole world, the unforeseeable effects of climate change and the dominant industrialisation of the production of food and its consequences for health. We will also try to understand, for the time being only from the theoretical point of view (due to a lack of empirical data), whether any of these phenomena can leave a trace on our epigenetic, rather than on our genetic, make-up. A whole chapter will be devoted to this aspect which today is central in medicine.

Three Exemplary Cases: Nauru, Greece and Bangladesh

Obesity in Nauru: A Story of Early Globalisation

Obesity is a dramatic reality in the Pacific islands, where very high rates of diabetes are also recorded. The changes in style of diet have certainly played a role, but the story is more complex than that. One almost textbook example is the small island of Nauru, with its only 10,000 inhabitants. The island became unexpectedly rich during the 1970s, thanks to the discovery of an enormous deposit of guano which was exploited to sell phosphates, used as fertilisers. The per capita income of the islanders increased dizzyingly, becoming one of the highest in the world by the end of the decade.

The increase in wealth was associated with many major social changes, whilst the excessive mining of phosphates was translated into a loss of farmable land. The consequence of all this was that the islanders' diet underwent a radical change, and the traditional consumption of fish and vegetables was replaced by a Western diet based on imported products. In addition, the islanders, having stopped farming and fishing, adapted to a sedentary lifestyle.

It did not take long for the consequences on health to appear. In 1975, the prevalence of diabetes had exceeded 30%, and in 2007, Nauru still had one of the highest rates of diabetes in the world, according to the data of the International Diabetes Federation. Unfortunately, the rapid exhaustion of the reserves of guano and a very poor financial management led Nauru to bankruptcy, and today the inhabitants have to cope with an epidemic of obesity and diabetes from a poor country's perspective. Three-quarters of the hospital beds are occupied by diabetic patients or with complications due to diabetes and there are only ten doctors on the island.

The story of Nauru is exemplary of how globalisation can affect many low-income countries. There are all the elements of an announced tragedy: the increase in wealth linked to exhaustible primary resources, the tendency to make risky financial investments, the destruction of the traditional economy (agriculture and fishing), the inclination to spend money on superfluous consumer goods and purely of prestige (such as powerful cars on an island where there is only one road), the drastic reduction in physical activity and the import of industrially processed foods replacing traditional food. It must therefore not come as a surprise that in this sort of "laboratory" the effects on the state of health appeared more quickly than elsewhere.

With regard to the import of industrially processed food, the term "spam" comes from a brand of tinned meat made with meat off-cuts, which was widespread in the Pacific (in particular on Hawaii) in the period after the Second World War, and which contributed to the epidemic of obesity that appeared there. The term then took on a metaphorical meaning in the world of communication, apparently due to a famous Monty Python programme in which Spam was a constant but unwanted ingredient in every dish in a "low-cost" restaurant.

Greece: The Health Effects of Economic Crisis

Greece is an involuntary "workshop" for studying the effects of the recent economic crisis, also due to the speed with which it appeared: between 2008 and 2010, male unemployment rose from 6.6% to 26.6%, and youth unemployment from 19% to 40%, whilst industrial production dropped by 8%.

In 2011, the editor of "The Lancet", Richard Horton, launched an appeal for the publication of data on the consequences of the economic disaster on health and the answers came in a series of articles in 2013 and 2014 (Karanikolos et al. 2013; Kentikelenis et al. 2014) based on current data available at the European Community and on bespoke investigations.

Thanks to the information collected in two random samples of 12,346 and 15,045 people, interviewed respectively in 2007 and 2009, and using the reports from medical institutions and non-government organisations, the authors of a paper (Kentikelenis et al. 2011; see also Karanikolos et al. 2013) recorded an increase in recourse to hospitals of 24%, although this coincided with a 40% cut in hospital budgets; an increase of 14% of those who described their health as "bad" or "very bad"; and an increase in suicides of 17% and a soar in violence and murders. Even more worrying are the increased use of heroin (+ 20% in 2009) and the number of HIV-positive people (+ 52% in 2011 compared to 2010), as well as sharp growth of other infections in the first 7 months of 2011.

Although in Greece the trend of many of the health indicators is negative, the crisis has also had positive effects, such as a reduction in the consumption of alcohol and drunken driving, likely due to there being less money available.

According to another study, although from the start of the crisis in Greece the mortality rates had decreased overall, from 2011 the mortality of those over 55 years of age increased and one-third of this increase has been attributed to the austerity measures and more specifically to reduced access to health care (Vlachadis et al. 2014). The mortality rate has not increased (not even in particular age groups) in any Western country since the early twentieth century; the only exception was Russia after the collapse of the Soviet Union. However, how legitimate it is to attribute the changes in Greece to the austerity policies is not clearly documented.

There are other studies that suggest a far less negative result than that described by Vlachadis and his colleagues. For example, a paper used data from the European Union Statistics of Income and Living Conditions to compare trends in self-rated health in Greece and Ireland before and after the crisis with trends in a "control" population (Poland) that did not experience a recession (and had health trends comparable to both countries before the crisis). There was no significant change in self-rated health in Greece or Ireland following the onset of the financial crisis. However, a comparison with the control population suggested an increase in the frequency of poor health in Greece, with the effects most pronounced for older individuals and those living in high-density areas. There was apparently no effect of the financial crisis on the frequency of poor self-rated health in Ireland. The comparison is a very early one but leads to an interesting consideration: Ireland was able to get through the crisis without cutting pensions and other segments of the Welfare State in spite of the strong decline in salaries, which might explain the lower impact of the crisis particularly amongst the elderly population. Another interesting observation from the political point of view concerns Iceland: whilst rejecting the economic orthodoxy that advocated austerity, this country invested in social protection instead, apparently with very few adverse health effects of the crisis (Karanikolos et al. 2013).

Evidence is expected to continuously accrue on the impact of the economic crisis and austerity in Greece. Recently, an article in The Lancet reported that funding of public hospitals has been cut by more than 50% since 2009. Public health spending shrank to 4% of GDP by the end of 2015, compared with an European Union average of 6.9%. The impact on health care seems to be devastating according to this paper (Karamanoli 2015).

The economic crisis and depression may have long-lasting impacts. For example, one common observation is that people tend to stop making investments for their health which in the long term would have had great benefits, such as dental care and eating fresh fruit and vegetables.

The Case of Bangladesh

Bangladesh is often quoted as a country-laboratory because it is characterised by at least three changes typical of globalisation and linked to health: massive migration (mainly male) to wealthy countries such as Dubai and Qatar, vulnerability with regard to climate change and the recent spectacular reduction of infant mortality. Bangladesh is a "laboratory" for studies on global health also due to the demographic peculiarities, as it has half the inhabitants of the United States in an area equal to that of Florida, i.e. an extraordinary population density.

The first two phenomena (migration and climate change) correspond to the concept of global health—as we have defined it—because they are changes which transcend national borders and need to be tackled by joint international action. Migration has significant consequences for the work of doctors, nurses and public healthcare specialists in the developed countries: migrants, who in increasing numbers are fleeing from poverty, war or the effects of climate change (such as drought or flooding), have specific health problems, which are often difficult to identify and diagnose. The consequences of migration on mental health are particularly important. The high frequency of anxiety and depression amongst migrants can be attributed—along with other reasons—to their difficult economic conditions, to the awareness of the expectations generated in families by migration and to the cultural differences with the host country. A systematic review of literature shows rates of depression equal to 20% amongst migrant workers and 44% amongst refugees, which are very high values if compared with the average rates of the population in general (Lindert et al. 2009). The number of migrants in the world has increased from 150 million in 2000 to 244 millions in 2015 (1 every 33 people).

As far as climate change is concerned, according to the fifth report of the Intergovernmental Panel on Climate Change (IPCC) there is a high probability (or high confidence, according to the terminology of the IPCC) that the marine and coastline ecosystems of southern and south-eastern Asia will be affected by an increase in the level of the sea and that more than one million individuals will be at a risk of flooding in the coming decades. Climate change will also probably have serious consequences on agriculture and on the quality of water and, overall, massive effects on the state of health.

The direct effects on health (which appear in a few hours or days) are linked to the physical risks due to the floods, cyclones and some acute-onset infectious diseases such as cholera. The indirect effects (medium- and long-term, measurable in weeks, months or years) are due mainly to changes in the level of productivity of agriculture and the quality of food and to migration by large sections of populations, as a consequence of the scarcity of water or conflicts for possession of the land. The causal relationship of indirect effects with climate change is less obvious than for the direct effects.

Communicable diseases are amongst those most sensitive to climate change. The risk of malaria is related to the variability of the climate, exasperated in recent years by the cycles of "El Niño" in Asia, Africa and South America. Cholera epidemics are also favoured by changes in the quality and temperature of water, which increase the proliferation of the vibrio (Lara et al. 2009). Shrimp farming, which requires high levels of salt in the water, has developed rapidly in Bangladesh and has become a flourishing industry for export, but further worsens the ecological situation as it means that freshwater (in the fields previously used to grow rice) is replaced by briny water which facilitates the proliferation of the vibrio. As Ronald Labonté, an expert on the effects of globalisation, has said, "Today shrimps are competing with men" (Labonté et al. 2009).

The difficulty of doing research in real scenarios of climate change is illustrated precisely by the example of the intrusion of saltwater in the coastline areas of Bangladesh. In recent decades, the salinity of the surface and deep waters has reached levels never previously observed. Seawater has penetrated for over 100 km through tributary canals of large rivers, and now the problem of salinisation of freshwater affects millions of people and 800,000 hectares of arable land. The problem arouses serious concern as severe effects of salinisation have been reported in that area, in particular an increase in the frequency of cases of hypertension and gestosis (pre-eclampsia) during pregnancy (Vineis and Khan 2012; Khan et al. 2011; see also the film at http://tinyurl.com/phtte6q).

Although many of the factors are specific to Bangladesh, climate change is obviously the consequence of global changes. The Bangladeshi contribution to the emissions of CO_2 is almost irrelevant with respect to that of developed nations, but this country could be one of the first and main victims of climate change.

Coming to the good news, the rapid decrease in infant mortality has made Bangladesh an undisputed leader although a rapid improvement has also been observed in other low-income countries such as India.[1] The ambitious Millennium Development Goals (MDG), established by the United Nations in 2000, which aimed at obtaining significant successes in the state of health of poor countries by 2015, have been partly achieved, especially regarding infant mortality (fourth goal) and maternal mortality (fifth goal). In Bangladesh, great successes were obtained from as early as its independence (1971) thanks to skilfully coordinated local actions. Today, the number of children per woman and infant mortality are comparable to those of many affluent countries.

These examples from a "model" country show how crises and improvements live side by side in the era of globalisation.

---

[1] For this I recommend the interesting video by Hans Rosling available on "Gapminder": http://tinyurl.com/mdpl33r

3

Food

One of the most remarkable phenomena of our times is the alarming epidemic of obesity and diabetes (diabesity) which has affected almost the whole world. In 2005, some researchers even forecast that this epidemic could reduce the life expectancy of the Americans, though this has not materialised. The alarm must be taken with caution, and, instead of predicting a reduction in life expectancy, it would perhaps be more prudent to anticipate major instability in the system, especially to the detriment of the lower socio-economic groups.

The Two Faces of Nutritional Problems

We are going through the first period in the history of humanity when there are more people who are overweight than suffering from malnutrition. At the same time, however, malnutrition has not been completely defeated. The 1990s were a fortunate period for the fight against hunger (with a significant reduction in the number of people suffering from malnutrition) unlike the first decade of the new millennium when this success slowed down. According to the FAO (Food and Agriculture Organization), overall in 2012 there were more than 800 million chronically malnourished individuals in the world, 9 million of whom were in the more developed countries (http://www.fao.org/hunger/key-messages/en/; Hawkes et al. 2009). One even more widespread problem is the deficiency of micronutrients such as vitamins or folate, a phenomenon which affects many millions of inhabitants in the poorest countries.

© Springer International Publishing AG 2017
P. Vineis, *Health Without Borders*, DOI 10.1007/978-3-319-52446-7_3

However, the other face of nutritional problems are the approximately 1.6 billion people who are overweight, of whom at least 400 million are obese (and the number is rapidly increasing). Obesity is closely associated with the risk of diabetes (the phenomenon is designated by the neologism "diabesity"): the number of diabetics is destined to grow from about 170 million in 2000 to at least 366 million in 2030 (with fluctuations in the estimates), except for extraordinary but improbable successes in prevention (Wild et al. 2004; Hawkes et al. 2009). This is why there is talk of the double burden of malnourishment and obesity in low-income countries (we prefer to use this definition rather than the more common "developing countries" because the degree of development is difficult to define and is extremely variable).

Whilst the causes of malnutrition can easily be identified in the structural poverty of many geographical areas and of many families, the causes of obesity and diabetes are to a great extent unknown. Some facts are clear: in the United Kingdom, the consumption of refined sugars and fats has increased from five to ten times in the past two centuries, whilst the consumption of fibre-rich cereals has been drastically reduced. However, this kind of change has not been ascertained as the only cause of the obesity epidemic. Clearly, also a reduction in physical activity has played a central role (though the relationship between physical activity and overweight is still controversial in terms of causality: see Richmond et al. 2014). Concerning the latter, we need to see the positive side of it: much of the reduction in high-income countries is accounted for by domestic and workplace "labour saving" innovations—which have improved lifestyles particularly for women and increased safety in the home and workplaces.

To combat obesity we need to tackle both excessive intake of poor quality food and the lack of physical exercise. However, this fight cannot be exclusively based on appeals to individual responsibility but requires structural changes. One of the greatest limits in preventive measures lies in placing the responsibility on the single individual, their greed or their laziness (some of the characteristics stigmatised in the character of Homer Simpson, lounging in an armchair in front of the TV with a beer in one hand and a doughnut in the other). This aspect is certainly real and important, but the more remote determinants of obesity are still little explored: for example, the "built environment", i.e. the speed and nature of urbanisation. This proceeds much faster in poor countries than in rich ones, has particularly important effects for low-income families and is accompanied by changes—such as access to cheap food of low quality, or the reduction of physical exercise—which influence obesity.

Without reaching the emblematic case of Nauru, all over the world millions of families have replaced a family economy based on manual labour and self-consumption linked to the land by urbanised lifestyles and being dependent on purchasing cheap food.

The Global Food Industry

The relationship between the developments of the global food industry, with its strategic decisions and the consequences on health, is still little explored. A special issue of the journal "PloS Medicine" published in 2012 drew attention to the fact that there is still very little research on the strategies of the food industry and on how they condition the health of millions, or even billions of people. The authors of the editorial recall that the effects of tobacco were recognised late and were denied for a long time by the industry; it is only now, more than half a century after the proof of the link with cancer, that incisive policies of prevention have been started in the developed nations.

The food industry follows strategies which are not very different from those used in the past by the tobacco industry, such as massive investments in advertising, often aimed at subgroups of the population, or corruption of researchers (a phenomenon of which we probably only know the tip of the iceberg; see below). Mortality by lung cancer in the United States between 1960 and 2010 doubled (from 24 to 48 deaths every 100,000 per year); the frequency of overweight and obesity (BMI [body mass index] > 25) in the same years went from 45% to 68%, a rate of growth which shows no signs of decreasing. Notice that those for obesity are percentages, not cases per 100,000: the overall numerical impact is much more marked with respect to the data referred to lung cancer (even though obviously obesity is a far less serious condition than cancer).

Macroscopic changes have taken place and are taking place in the production and distribution of food. The liberalisation of trade has facilitated the so-called vertical integration of the TFP (transnational food processors): the food industry can cover all the segments, from harvest to distribution, in an integrated system of a highly industrialised type, with enormous contractual benefits and a reduction of costs at all levels. In the United States, the degree of concentration (the market share covered by the first four corporations in the sector) for the retail food trade went from 24% in 1996 to 46% in 2003. And it is ironical that the current "universal equivalent" is no longer the dollar but the "Big Mac" (Big Mac Index used by the economists).

In England most people go shopping at Tesco, Sainsbury's and in a few other large supermarkets characterised by scale economies. Cutting costs started from the personnel: according to the lesson of Ryanair, costs are cut by having the customer do the work, and personnel has been replaced by self-service cash desks. The economist Atkinson has written interesting pages on the desertification of human relationships—in addition to the impact on occupation—from the progressive substitution of humans with machines not only in factories but now in commercial enterprises (Atkinson 2015).

Alongside the TFPs, the other great innovation in the past decades has been the FDI, foreign direct investments, i.e. the fact that transnational corporations (which answer to the shareholders and not to the clients) invest financially in food, considered a consumer good like any other. The investment returns much more if the food is not simply a product of the land, grown and transported to the consumer, but if it is transformed: indeed, the more it is transformed, like fizzy drinks and packaged food, the greater the economic return. The missing explanation of the obesity epidemic may lie in exactly this aspect. English or American supermarkets—such as Tesco and Sainsbury's in the UK—compete on the ratio between price and attractiveness of packaged foods targeted at low-income groups: the widespread advertising for increasingly new types of hamburgers are an example. These operations of diversification are carried out by the large chains, increasing their profits. Small food shops are being reduced, at least in the large Western cities, to boutiques of genuine but very expensive products that only the upper middle classes can afford.

What has been said about retail also applies to restaurants: street food and fast-food outlets (mostly linked to major chains) are increasingly widespread, in particular in Eastern Europe, Asia and Latin America. In Vietnam, China and Indonesia, the estimated growth is respectively of 11, 10 and 8% per annum. The proportion of calories consumed from food distributed by fast-food outlets in the United States went from 10% in 1977 to 21% in 1996 (Moodie et al. 2013; Hawkes et al. 2009). One collateral but not negligible aspect is that almost all pre-packaged foods are wrapped in cellophane or in plastic and contribute to the production of pollutant material which is difficult to dispose of.

A very nice photographic essay on the rapid changes in food markets in Thailand between 1988 and 2006—including the use of plastics—has been published by the International Journal of Epidemiology (Dixon et al. 2007, and www.ije.oxfordjournals.org). In this particular case (and probably many others), the message from the authors is mixed and essentially related to the income of customers. According to the authors, industrialisation has brought

Thais more pre-processed foods and more dietary sugar, fat and animal products. On the other side, "The increasing diversity of market foods, particularly green vegetables and herbs, is health promoting as long as residents have sufficient income to participate in the growing market economy."

What does the future hold in store for us? Possibly more and more industrial and (badly) processed food? The structural adjustment programmes and the liberalisation of trade can only increase the strength and the spread of the great TFP. If the WTO opposes limiting the sale of cigarettes, it is unlikely that it will step down in the face of the criticism of the food industry; consider that at a worldwide level, transactions related to food account for 11% of the total, which is a greater percentage than for fuel. We will return to the resistance to policies of limiting the spread of junk food in the last chapter.

The phenomenon of the globalisation of the food industry has been particularly intense in some structurally weaker areas. The Pacific islands represented a natural laboratory for the policies of selling pre-packaged food by the United States and Australia, in particular regarding tinned meat (often obtained from the less noble parts of sheep and turkeys). Globalisation has also caused a drastic reduction in the consumption of fresh fish on the same islands, in favour of the consumption of tinned fish, following the sale of fishing concessions to the Japanese.

One powerful motor of the development of the transnational food industry—as in all other industrial sectors—is advertising. The global expenditure on food advertising has increased from $216 billion in 1980 to 512 in 2004, and in the United States the food industry spends more than any other industry on advertising (Moodie et al. 2013; Hawkes et al. 2009). Much of this advertising, very present on American TV channels, promotes products with a high content of calories and fats: pizzas groaning under cheese, hamburgers with triple layers of meat, etc. In 2013, in the Christmas period, in the London Underground Sainsbury's advertised the whole Christmas lunch (turkey included) in a single hamburger for £1.99: how depressing!

One of the most aggressive companies in advertising campaigns is Frito-Lay, which in Thailand more than doubled its investments in advertising between 1999 and 2003: in the same period, the consumption of snacks by children grew by 30% in the country. Frito-Lay (a division of Pepsi-Cola) is also to be remembered for one of the many episodes of conflict of interest involving medical research. A group of researchers recently published a review that negates the proof that acrylamide, a substance which is produced in frying (e.g. of crisps), is carcinogenic. Beyond the evaluation of the evidence in itself, some facts are striking: the research was sponsored by Frito-Lay and the senior author of the article worked for many years for the IARC, the International

Agency for Research on Cancer, before joining private industry, for which he is now engaged in systematically revising the evidence of carcinogenicity collected by the same agency he previously worked for. We will return to conflicts of interest in Chaps 7 and 8.

I have mentioned the fact that the recession, but also the liberal solutions that have been proposed, such as the structural adjustment programmes, can entail rapid deterioration in the state of health in certain parts of the world. Surprising as it may seem, one of the most prestigious international journals of medicine, the "New England Journal of Medicine" (NEJM), published an article in 2005 which hypothesised a decline in life expectancy in the United States in the twenty-first century (Olshansky et al. 2005). This decline would be mainly driven by obesity, with its associated burden of cardiovascular diseases and diabetes. The increase in life expectancy—constant since 1850 in high-income countries—has undergone a deceleration in the past three decades, in particular in the United States. The probability of dying early can be up to four times higher for morbidly obese people, i.e. with a body mass index (BMI) >40, aged 18–30, and a diagnosis of diabetes at an early age can reduce life expectancy by no less than 13 years (Olshansky et al. 2005). Overall, extreme obesity (BMI > 40) reduces life expectancy by a period of between 5 and 20 years.

According to the estimates of the authors of the article in the NEJM, if there is no turnaround soon in the obesity epidemic, the overall life expectancy of Americans could be reduced for the first time since 1850. In fact, this may have happened earlier than expected. The 2015 Nobel Prize winner for Economics, Angus Deaton has published mortality statistics suggesting that there would be already a decline in life expectancy amongst white Americans (Case and Deaton 2015). A midlife mortality reversal has been observed amongst middle-age white non-Hispanics, whilst blacks and Hispanics continued to see mortality rates fall. This increase for whites was largely accounted for by increasing death rates from drug and alcohol poisonings, suicide and chronic liver diseases and cirrhosis (but apparently not obesity-related illness); those with less education saw the most marked increases. However, the publication of the paper has been followed by criticisms, including potential bias in their observations, so that "the jury is still out".

A slowdown in the increase of life expectancy has also been observed in Japan, in Okinawa, as a consequence of obesity and cardiovascular diseases. The roots of the latter observation, however, are more complex and are linked to the low birth weight of a whole generation. We will return to this point in the chapter on epigenetics.

The analogy between the obesity epidemic and that due to cigarette smoke—which has caused not only millions of cases of lung cancer but also millions of cases of cardiovascular and respiratory disease in the past decades—deserves being looked at more closely in all its dimensions.

The world is finding it very hard to be free of the consumption of tobacco, at political and individual level, and the great spread of diseases linked to smoking, which began in the early decades of the last century in the rich countries, has not yet stopped and is also extending to emerging countries. We have to avoid that what happened with smoking is repeated with obesity and diabetes, possibly as a consequence of that great rationalisation and concentration in the production and distribution of food which is appearing in coincidence with the global economic crisis.

4

Climate Change

This chapter discusses what we know today about climate change, the uncertainties accompanying our knowledge and the consequences that climate change may have on health through different mechanisms. Climate change seemed inconceivable a few decades ago and is a clear example of how human influence can have a global impact on our planet. The main consequences on the state of health will, in all likelihood, appear through the scarcity of (good quality) water and through the changes in the distribution of communicable diseases such as malaria.

The Fifth Report by the IPCC: The Effects on Health

The chapter on health (Working Group 2; http://www.ipcc-wg2.gov/index.html) in the fifth report by the IPCC contains much data which brings further evidence on the importance and speed of climate change. For example, 13 of the 14 hottest years since temperatures began to be recorded have been in the twenty-first century. The report, like many other authoritative sources, draws attention to the most dramatic consequences that there will be in forthcoming years.

The first is certainly the insufficient availability of water resources of good quality for everyone. Although there is no evidence that at global level the speed at which surface and deep water is used up has changed, this is expected

© Springer International Publishing AG 2017
P. Vineis, *Health Without Borders*, DOI 10.1007/978-3-319-52446-7_4

to happen in arid subtropical areas in a few years' time. In the long term, a serious shortage of water is expected due to at least three concomitant phenomena: the increased demand, changes in rainfall and the melting of glaciers. With respect to current availability, it is estimated that each increase of 1 °C in the temperature will cause a drop of 20% in renewable water resources for a further 7% of the population (IPCC, https://www.ipcc.ch/report/ar5/). In addition, the increase in temperature will also lead to an increase in the sediments in sources, drought will cause a reduced dilution of chemical contaminants and flooding will lead to a cyclic collapse of waste disposal systems.

To give just one example, amongst many, of the long-term implications that water scarcity and drought may have, a paper in the Proceedings of the National Academy of Sciences has shown that before the Syrian uprising that began in 2011, the greater Fertile Crescent experienced the most severe drought in the records. For Syria the drought had a catalytic effect, contributing to political unrest. Precipitation changes in Syria were linked to rising mean sea-level pressure in the Eastern Mediterranean. According to the analysts, a drought of the severity and duration of the recent Syrian drought has become more than twice as likely as a consequence of human interference in the climate system (Kelley et al. 2015).

If no measures of mitigation and adaptation are taken, by the year 2100 it is expected that millions of people will suffer the consequences of flooding, whilst coastal erosion will provoke a loss of arable land. According to the report, the effects of climate change on food production depend on complex interactions between levels of CO_2, nitrogen and ozone, temperature, availability of water and extreme climatic events, all of which are phenomena the extent of which is difficult to predict. Although positive effects have also been observed on agricultural production due to climate change, to date these have been exceeded by the negative effects, especially in the case of maize.

According to estimates, harvests overall could be reduced by up to 2% by decade in this century, whilst until 2050 demand will increase by 14% by decade (https://www.ipcc.ch/report/ar5/). Food production is greatly affected by strong climate differences, and if there are extreme temperature increases (more than 4 °C), the damage in this sector—which is now limited to temperate areas—will be perceived at all latitudes. Climate change is a threat multiplier for the poorest social groups, also due to the increase in food prices. It is likely that this will not be a problem only in low-income countries, as it is today, but new areas of poverty will be created in high-income countries, where the levels of inequality are on the rise.

The consequences of climate change will be different in the different continents. In Europe, the IPCC report estimates as "highly probable" an increase in the number of people affected by flooding in river valleys and along the coasts, with consequent huge economic losses. At the same time, there will be a reduction in water resources together with an increase in demand. In other continents, there may be important consequences on diseases transmitted by vectors, as suggested by an article in "Science" on the diffusion of malaria at high altitudes in Ethiopia and Colombia (Siraj et al. 2014).

As far as the preventive and protective measures are concerned, it is fairly obvious that the strategies of "mitigation"—i.e. of reducing levels of CO_2— can produce results only in the long and very long term, whilst measures of physical and biological "adaptation" to the current situation can aspire to having faster effects. However, no evaluation of the efficacy of the measures of adaptation is yet available, nor are there any indices of a change in territorial policies according to climate change (e.g. in territorial planning outside built-up areas or in the reduction of over-development). Adaptation can be extremely complex and expensive (from building river barriers, such as the one on the Thames, to the other infrastructures necessary to protect ports and coastal areas). In some cases, however, adopting very simple measures can be effective, such as mangroves, algae and salt marshes which, as well as protecting the coast, absorb CO_2.

In conclusion, the fifth report of the IPCC reinforces the previous signals aimed at asking for urgent action to avoid phenomena, which are no longer reversible, being triggered off.

Climate Change and Communicable Diseases

Water is strongly related to the spread of communicable diseases. Every inhabitant of the planet ought to have access to a sufficient quantity of water of good quality, which is not contaminated and not stagnant. By 2025, about half of the world's population will be living in conditions of extreme scarcity of water, and in many parts of the world, the quality of water is deteriorating. About half of the humid areas, with their characteristic flora and fauna, have already been lost, whilst 70% of the reserves available are used today for irrigation. The fact that there is a very strong component of social injustice cannot be ignored, both because it is the poor who do not have access to good quality water and because it is the wealthy who are responsible for wasting it and for "futile" uses such as irrigating golf courses in arid areas.

The change of the local microclimate brings with it a vast set of consequences due to the spread of insects which act as vectors, and the temperature changes facilitate or inhibit the proliferation of bacterial or parasite species. The ways that communicable diseases can be spread are multiple and are usually divided into four simple categories.

Diseases *"transmitted by water"* are those by oro-faecal transmission, such as cholera or various forms of diarrhoea. Cholera is still endemic in many countries, in particular in Bangladesh where, not surprisingly, the most prestigious hospital and research centre, the ICDDR, B (International Centre for Diarrhoeal Disease Research, Bangladesh), is specialised in this disease. The increase in the temperature of the sea and the inland waters facilitates the proliferation of the vibrio and cholera is changing its geographical distribution.

Water-based diseases (*"based on water"* but not directly transmitted by it) are, on the other hand, those where a parasite spends part of its life cycle in water, as in the case of schistosomiasis. There are also signs of the spread of this disease outside the endemic areas, for example in some parts of China, and there are future projections about schistosomiasis that arouse concern for many low-income countries. Specific types of Schistosoma are carcinogenic parasites and cause bladder and liver cancer in many poor countries.

Again according to the traditional classification, *water-washed diseases* are those in which the causal agents are usually eliminated if rules of elementary hygiene are followed; examples are scabies and trachoma. In this case, the crucial problem is the availability of water for washing, and therefore a reason for concern is the desertification of extensive areas of the planet.

Lastly, *water-related diseases* are due to the fact that the vector of a parasite has a life cycle that includes water. The best-known example is malaria, linked to the cycle of the Anopheles and the presence of stagnant water. Malaria is perhaps the disease that has been most studied in relation to climate change and there is evidence of its diffusion outside the areas where it is endemic (whilst in other areas it is being reduced). The subject was tackled in an issue of "Science" (Siraj et al. 2014) in which the effects of the temperature changes in the highlands of Colombia and Ethiopia are described. The article concludes that climate change will lead to a spread of malaria in the densely populated highlands of Africa and South America.

It is important to note that the change in the distribution of communicable diseases following climate change is a phenomenon which will not exclusively concern low-income countries. In the past few years, we have been seeing the appearance in Europe of infectious diseases which previously did not exist or were unknown—for example chikungunya or the Zika virus, both transmitted by the Aedes mosquito—which are linked not only to migration and transcontinental travel but also to climate change, or the interaction of the two factors. This latter problem, that of the interactions, should be of interest and concern to epidemiologists in particular, as the concomitant phenomena of mass migration and climate change can induce unforeseeable and even uncontrollable multiplying effects (see http://tinyurl.com/q4bqb9r).

Causality and Climate Change

Over the years, epidemiology has produced a body of knowledge that now allows transmitting its methods in a rigorous and standardised way. There is now also a certain degree of consensus on the concept of causality, i.e. on how to attribute the effects to the causes. However, it is difficult to venture into the terrain of new disciplines which correspond to the rapid changes under way in the world and specifically "global health" and climate change. This new discipline raises serious problems of "causal attribution", because here the conceptual difficulties of epidemiology and public health are multiplied: (a) experimentation, as carried out for example with drugs, is not possible; (b) the reconstruction of the causal chain is, to a great extent, incomplete and obscure; (c) the possible indirect effects are so many and so different that untangling the plausible causal chains from the less plausible ones is problematic; and (d) the usual recourse to molecular mechanisms supporting causality, as for the causes of cancer, is not easily applied to climate change.

The Italian philosopher Federica Russo has contributed to systematising the conception of causality in epidemiology with the theory known as Russo–Williamson (McKay Illari et al. 2011), according to which causality travels along the two lines of difference-making and the explanation based on molecular mechanisms. For example, if an epidemiological study describes an excess of angiosarcomas of the liver in workers exposed to a chemical substance ("difference-making"), the biological plausibility of the possibly causal relationship can be confirmed and reinforced by observations on the cells, on laboratory animals and on molecular markers in the blood of the exposed workers. Considered altogether, this knowledge about the mechanisms allows reconstructing the sequence of events that from exposure lead to the onset of cancer and therefore increases the causal plausibility of the observation in the exposed workers.

Going over the history of the theories of causality, we can see that they were initially developed from classic physics, through the study of closed (artificial) and deterministic systems: in this case the cause always gives rise to the effect. Physiology (the study of the normal functioning of living beings) is in some way inspired by physics by trying to establish closed experimental systems, in which the external cause is introduced by the experimenting agent who manipulates the conditions of exposure. The study of pathology has discovered the elusive nature of cause–effect relations, as a consequence of the high individual variability in response to pathogenic agents and the presence of mechanisms of repair and "biological reserve". The study of causal relationships in human populations implies probabilistic relationships but also keeps a reference to basic biology and molecular mechanisms, as maintained by the Russo–Williamson theory. However, climate change and the study of its consequences on health today need even more complex and sophisticated models, with marked differences compared to all the earlier ones: (a) they are not and cannot be closed causal models as in physiology since the climate is a very open system; (b) we still do not know in detail how the individual "biological reserve" (and repair capacity) appears in this case; and (c) the variables that interfere with the cause–effect relationship are multiple and experiments are not possible. Considering point (b), the "biological reserve" is the concept that is applied to, for example, the pulmonary functionality following the cumulative exposure to harmful agents, such as dust and smoke: the individual has a physiological reserve that allows him to meet the damage caused by the environment, but at a certain point the reserve is exhausted and the veritable pathology intervenes (such as chronic respiratory diseases). In the case of diseases linked to climate change, it is probable that the effects are not proportional to the intensity of the exposure but are non-linear. It is also probable that the biological reserve varies enormously depending on the subgroups of the population: typically, heatwaves cause far more deaths in groups of more vulnerable people such as the elderly and infants.

The degree of regularity of the observations is gradually weakened from physics to physiology, to pathology, to epidemiology and now to climate change: bodies fall according to a well-known, simple and regular law of physics; blood pressure is explained by physiology in a fairly clear and consolidated way, but the causes of hypertension are made up of far more complex and probabilistic "networks". Lastly, for phenomena such as the effects of climate change on health, it is almost impossible to isolate simple causal chains, distinguishing them from phenomena of perturbation or interference, and the mechanisms that mediate a possible causal relationship are often unknown or at least difficult to systematize. For example, on the one hand we can have a war triggered off by the drying up of land and the crisis of agriculture and on the other hand a migration in mass of species of fish traditionally caught by the inhabitants of an island. These are two completely different mechanisms which interact in turn with other surrounding conditions and which influence, through independent causal chains, the health of thousands of individuals.

Climate Change and Vulnerability

It is now a cliché to say that not everyone is equally vulnerable in the face of climate change. Inequality at worldwide level is macroscopic and is related to the asymmetrical distribution between those who pollute (the rich countries) and those who suffer the consequences (the poor countries). In this sense, the effects of climate change do not correspond to the trend with respect to what has been observed for the distribution of wealth and other indicators, for which today the differences inside the countries are often greater than those between countries (see for example Bourguignon 2012). The vulnerability to extreme climatic phenomena affects above all the poorer sections of the population: one glaring example of this was the flooding of New Orleans. At the same time, climate change continues to produce inside each country inequalities which are almost as glaring as those between countries.

In Bangladesh, where a group of researchers interviewed 700 residents in an area close to the coast, frequently subject to flooding (Brouwer et al. 2007), the poorest families were for various reasons more subject both to flooding and to its consequences: their homes were more fragile and built in places where others did not want to build; the possibilities of recovering after flooding were more limited and their preparedness was more fragmented. Access to medical care was also more limited. The flooding in the houses reached a level of more than 1 m in families with an income below $600 per annum and of 10 cm for those whose income was over $1200. The average annual income in the families that had adopted protective measures in advance was of $1500, against the $976 of those who had not; the damage amounted to $245 for the former and $391 for the latter: poverty often has a tendency to be self-aggravating.

The political lesson that can be drawn from this and other examples is clear, but the measures necessary are complex: reducing the vulnerability of the most disadvantaged in the face of climate change requires massive investments and on a large scale. In Bangladesh, it would mean for example moving vast sectors of the population and building shelters and more suitable homes with clean water.

Denialism

As the book by Oreskes and Conway (2010), effectively entitled *Merchants of Doubt*, shows, there are various ways to throw discredit on to a scientific theory, and one of these is to raise constant doubts on the validity of the evidence. Putting alongside the evidence in support of a certain theory other evidence (possibly inflated) that contradicts it is a common way to raise a smokescreen, cast doubt and divide public opinion; this technique has been used for active smoking and then for passive smoking, for acid rain, for ozone, for pesticides and recently for global warming. As the book shows, very often the scientists involved in deconstructing the evidence are the same, engaged on very different subjects, but to which they apply the same "smokescreen" methodology.

One of the most violent attacks on the theory of global warming was published on 27th January 2012 in the columns of the "Wall Street Journal", in an appeal signed by 16 scientists and entitled *No Need to Panic about Global Warming*. The six main points raised by the critics were answered by William Nordhaus, a Yale economist who has devoted years to the study of climate change. It is worth summarising here the arguments of the two fronts, to give an example of how some scientific controversies are played out today, and which often have less than ideal motives (Nordhaus 2012).

The first argument of the critics is that the planet is not warming up and in particular it has not warmed up in the past 20 years. The counter-argument of Nordhaus is that the critics are only looking at short-term fluctuations, and not in the period from 1920 to the present—a period when the temperature has increased by over 0.8 °C. By choosing only the past decades, they fall under an area of wide casual fluctuations, the general trend of which, however, follows that of the previous years.

The second point: the warming-up is alleged to be less than that predicted by the models. The answer is that the more serious scholars have considered different starting conditions, they have introduced various kinds of assumptions in the models, tested the sensitivity of the models to these assumptions, etc., but they have always obtained the same results.

The third point: the 16 attack the idea that CO_2 is an environmental contaminant, as it is not toxic for living organisms at the concentrations usually found. It is clear that this argument is ill-taken as we are not talking about the effects on humans or on other species but about the effects on the atmosphere.

The fourth point: the 16 critics maintain that they are living in a climate of a witch-hunt and that expressing dissent is particularly dangerous in terms of career and personal safety (they propose an analogy with the Russian geneticists under Stalin). Nordhaus argues that on the contrary, the predominant culture in the media puts dissent on centre stage and gives those who dissent far more space than they deserve on the basis of the scientific quality of their arguments.[1]

The fifth point, linked to the previous one: the scientists who maintain the theory of climate change enjoy academic and financial advantages. This argument is to be considered provocative if we are to think of all the energy and money invested by the tobacco industry to corrupt scientists and produce evidence contrary to the current data and the same was happening recently with climate change (as well documented in Oreskes and Conway 2010).

The sixth and last point: the 16 critics maintain that the economists advise against implementing policies aimed at reducing climate change in the coming decades in order not to hinder development, especially in times of crisis. Nordhaus is an economist and does not find it hard to show, as others have already done, that the negative effect on the economy of measures against global warming would be minimal, against the enormous benefits that there would be in the long term. According to one estimate, an effective policy of mitigation of climate change from now until 2030 would cost less than 3% of the global gross products; that is it would put development back by about 1 year (Andy Haines, personal communication). According to an authoritative economist, Dani Rodrik (2011), if a Tobin tax of 0.1% were to be introduced (i.e. if financial transactions were taxed by this amount), the earnings would be sufficient to start an incisive policy to reduce the effects of climate change. If one takes the co-benefits of reducing greenhouse gases and air pollution (in particular for health and life expectancy) into account, then there would be net economic benefits, according to calculations (West et al. 2013).

What is the greatest lesson that can be learned from this controversy? The importance of anchoring reasonings to a rigorous method: without a rigorous method to collect and interpret evidence, one theory is as good as the next one, because it is opinions that prevail. Unfortunately, questions of method are not very popular with the mass media, which "focus on the news" and often do not worry about distinguishing between instrumental and biased controversies and solid scientific data supported by a rigorous research method.

---

[1] This comment was still true in 2015, but not longer after the COP21 conference in Paris (December 2015).

5

The Environment

Whilst some consequences of globalisation are now widely studied, such as the epidemics of emerging or re-emerging infectious diseases, others are far less, for example the massive exposure, in low-income countries, to environmental contaminants such as asbestos, arsenic or atmospheric pollution. In this chapter we will be examining some important environmental effects of globalisation, which also have consequences on human health.

Aflatoxin

One of the most important environmental carcinogens of those that mainly affect (although not exclusively) the poor countries is aflatoxin. Aflatoxins are a class of toxic metabolites produced by some species of fungi such as Aspergillus flavus, which contaminate nuts and grains. Laboratory tests have shown that aflatoxins are carcinogenic in various animal species, in which they cause cancer of the liver, the main organ targeted by this family of contaminants. Liver cancer is one of the commonest tumours in the world, with a wide variability in geographical distribution: relatively rare in developed countries, its incidence is very high in Africa and in South-East Asia. In a study on 18,000 people in Shanghai, individuals with high levels of aflatoxins in urine together with an infection from the virus of hepatitis B had a 59 times greater risk of developing liver cancer, whereas in those who were not exposed to the virus, the effect of aflatoxin increased the risk by about five times (Ross et al. 1992). The number of people exposed to aflatoxins in the world has not been

© Springer International Publishing AG 2017
P. Vineis, *Health Without Borders*, DOI 10.1007/978-3-319-52446-7_5

measured, but it is estimated to be more than 5 million. If we take a risk increased by five times from those exposed (regardless of exposure to viruses), it follows that the number of liver cancers that can be prevented by abolishing exposure to the toxin would be at least 68,000, 80% of the 85,000 cases ascertained each year in the populations in low-income countries, exposed to the toxin.

It is probable that exposure to aflatoxins in poor countries will worsen due to climate change. Climatic factors foster the growth of fungi such as Aspergillus and the contamination of foodstuffs. The fungi that produce aflatoxins, as they are not very sensitive to climate changes, are better equipped to compete with other species, and high temperatures during the critical stages of maturation of the crops easily lead to contamination. Not only the changes of temperature but the growing unpredictability of the weather will also probably contribute to the increased contamination of food by aflatoxins. In tropical areas where aflatoxins are a constant threat, rain close to harvest time leads to particularly high levels of contamination, even though the populations exposed have traditionally coped by implementing strategies of adaptation. However, when the rhythm and quantity of rainfall changes, even traditional strategies become useless (Cotty and Jaime-Garcia 2007; Paterson and Lima 2010). An episode of acute mass intoxication, in 2004–2005 in Kenya, which caused the death of 125 people, was induced by a variation in rainfall, which resulted in a delay in harvesting maize and its unsuitable storage (Probst et al. 2007; Lewis et al. 2005).

Global climate change will lead, if counter-measures are not taken, to aflatoxins spreading even to latitudes where for the time being they are unknown—perhaps even to Europe—and to an increase in the number of people exposed, together with a possible increase in cases of liver cancer.

Arsenic in Water and Other Environmental Exposures in Poor Countries

At least 137 million people in the world, 70 million of whom live in the plain of the Padma-Meghna, in Bangladesh, and in the areas bordering with India, are exposed to arsenic through drinking water. In most of these areas, the problem did not exist until the 1970s, when deep wells were dug to provide water uncontaminated by bacteria and parasites: unfortunately, the rocks turned out to be very rich in arsenic which was released by drilling.

Chronic exposure to arsenic causes skin cancer and tumours of the internal organs (lung, bladder and kidneys) with a direct relationship between the dose and the risk. One study has estimated that exposure to arsenic in Bangladesh is responsible for the doubling of death rates by lung and bladder cancers (Chen and Ahsan 2004).

Another important problem, which is also typical of the poorest areas in the planet, is that of illegal landfills, which cause exposure to carcinogenic agents such as dioxins, PCB, arsenic, cadmium, nickel, polycyclic aromatic hydro-carbons and solvents (including benzene). In poor countries, illegal landfills are more common, more frequently contaminated by toxic chemical substances and closer to homes than in rich countries. The FAO has estimated that 120,000 tons of toxic waste, mainly obsolete pesticides, have been dumped illegally in Africa.

Another cause of environmental tumours in poor countries is exposure to smoke from biomass. Biomass, a general term to indicate organic substances such as wood, agricultural residue and animal excrements, is widely used as a fuel for heating and cooking. The combustion of these materials gives rise to polycyclic aromatic hydrocarbons and other carcinogenic substances. It has been estimated (Vineis and Xun 2009) that each year tens of thousands of lung cancers could be prevented in low-income countries by reducing the exposure to the products of combustion of biomass, for example through simple actions increasing ventilation in homes.

Urban Sprawl, the Built Environment and Health

It does not appear that the current epidemic of obesity and diabetes is explained in full by changes in dietary habits and physical exercise, considered singly. The role of changes in food consumption and energy expenditure can in part be understood through what are known as non-linear interactions, a key component in "catastrophe theory". The latter means that the overall effect of several causes is greatly superior to the sum of their separate effects. Non-linear interactions are essential to understand and forecast macroscopic phenomena such as credit crises and the collapse of financial markets, and they are also effective in explaining health phenomena.

One concept linked to non-linear interactions in the obesity epidemic is that of the built environment, defined as all those buildings, spaces and products that are created, or modified, by humans, which occupy the envi-ronment in various ways: housing, public buildings, factories, motorways, parks, commercial areas, transport systems, etc. (http://tinyurl.com/l292sg4; Lovasi 2009). Suburban areas are growing increasingly faster and often in an uncontrolled way (the "urban sprawl") and the consequences for health are of a non-linear type. Old cities—such as Perugia in Italy—were limited and had well-defined borders that separated them from the countryside. Homes, shops and productive activities were close to one another and walking was almost the only way to go from place to place. Today the city has expanded far beyond the traditional limits and the new town-planning design is dominated by at least two essential factors: large shopping centres and the use of the car. The shopping centres are places that offer all possible services and are therefore used in lieu of the old square, church and even one's own home. Even when distances are modest, in the new conception of town planning, the conformation of the town is such that in many areas using a car is indispensable.

It seems obvious that urban sprawl contributes to an increase in obesity through the interaction of its different characteristics (a sedentary lifestyle, consumption of industrial food, etc.). The effects of the built environment have been studied in New York (Rundle et al. 2007), using the methodology of geo-referencing (GIS, geographic information system). The authors have built up indicators that reflect the modes of urbanisation, for example by defining an index of "walkability" based on the urban design and on variables such as the population density: the index measures how easily pedestrians can use a certain part of the city, including taking safety into account. One of the concepts associated with the built environment is the so-called broken window syndrome: it has been proven that urban decay (expressed symbolically by broken windows in buildings, but also by litter, abandoned cars, the number of stray dogs, etc.) is associated with a lower degree of safety and therefore lesser "walkability". It is also a phenomenon that is self-fuelling, because degraded areas enjoy less respect and are further damaged by acts of vandalism and avoided by normal passers-by. Rundle and colleagues also examined the density of green areas and parks and the availability of fresh food in the different areas and, in the last place, compared their indicators with the rate of obesity amongst the residents in the areas examined, finding a relationship with obesity which was independent of the social status of the individuals. It is likely we will see a burgeoning of research similar to that conducted by Rundle, thanks to the development of new technologies to quantify urbanisation using remote sensing. For example, night-time light intensity is being used in India as an urbanisation index, and also prediction of malaria transmission is allowed by remote sensing (Rogers et al. 2002).

Another phenomenon linked to globalisation that can be observed in large urban concentrations is the progressive marginalisation (geographical and social) of those who do not work or consume, for example in the slums of Mexico City (two-thirds of the megalopolis) which do not contribute to the functioning of the city as a business centre (Labonté et al. 2009). The policies of liberalisation and opening up to trade tend to attract sectors of the population with a high purchasing power towards the centre of cities or to privilege it for intensive tourist exploitation, excluding the poorer sections of society which end up by benefiting from fewer services.

The "liberalised" use of the territory leads to consequences which are even more serious for the environment, such as the outsourcing of toxic waste, i.e. giving this waste to be disposed of by specialised firms. In fact, a flourishing economy has grown up around the illegal management and export of waste (Clapp 2001).

The Microbiological Environment

I will only briefly mention microbiological threats, not because they are not serious or real, but because the mass media tend to deal with them more and they are better known than many of the other issues discussed in this book.

Resistance to antibiotics is emerging as one of the biggest problems in hospitals, but also amongst the patients of general practitioners. Sally Davis, the British Chief Medical Officer, in one of her recent reports, speaks of a potential catastrophe looming on the horizon. According to her report, in 20 years' time, we could be seeing deaths from minor operations if new antibiotics are not discovered. There is a real "discovery void", linked on the one hand to the extremely high costs of the development of new drugs and, on the other, to the reckless use of existing drugs. Very few new antibiotics have been developed in the past two decades, whereas in the same period a new infectious disease has been identified every year. In the words of Sally Davies, the WHO and the G8 ought to immediately take the worldwide spread of resistance to antibiotics very seriously. Healthcare personnel also require education and awareness-raising on the problem, in order to drastically reduce inappropriate prescriptions. Exactly how many are the latter is not known with any accuracy (e.g. the administration of broad-spectrum antibiotics to people with influenza without complications) but we are probably talking about millions of doses (in the United States one-third of patients receive a prescription for antibiotics from their General Practitioner).

Lastly, many are awaiting the tidal wave of the next flu pandemic, linked in all likelihood to the large poultry and pig farms all over the world, veritable reserves of rapidly mutating viruses that transcend the barriers of species and can become extremely dangerous. When this may happen is not known, and it is to be hoped that the measures of containment will be (relatively) effective, like those used to defeat recent epidemics. It is well known, but it is worth recalling, that the 1918 outbreak of influenza killed between 20 and 50 million people in the world, although the comparisons with today's situation are not really possible due to the enormous improvement in the general state of health, in living conditions and in diet since then, and above all thanks to the capacity to produce vaccines in (hopefully) short periods of time.

6

The Economic Crisis

In this chapter, we aim to outline—although very generally—the intersections between the economy and health and in particular the effects that economic globalisation could have on health. There is still very little research on the consequences on health of the crisis that began in 2008, but the question we are asking in this chapter is whether it is possible to draw a parallel between the causes of the recession, the ways it has been tackled (austerity) and the foreseeable changes in the state of health, in particular the growing instability and a progressively growing gap between rich and poor.

In the first chapter, we mentioned the hypothesis that, as in the economic context, health can also be the object of similar and rapid worsening, especially at local level, as we have seen in the very different cases of Greece and Nauru. The levels of health reached today by many countries—never observed before in the history of humanity—are the expression of investments by generations, but they may be compromised in the space of very few years or at least undergo great fluctuations.

What happened with the crisis that began in 2007–2008? After the fall of Communism and the unification of Germany, the development of the European Community was stepped up, as is well known, on a mainly monetary basis but with wide institutional shortcomings. In particular, a central bank was created, but not the possibility of issuing European bonds and neither a common fiscal policy, in the belief that politics would have remedied in the case of need. In the week after the bankruptcy of Lehman Brothers, the whole world of finance risked collapse and was kept alive artificially, to put it in the words of George Soros (2012).

© Springer International Publishing AG 2017
P. Vineis, *Health Without Borders*, DOI 10.1007/978-3-319-52446-7_6

In concrete terms, the governments initially saved the major banks (too big to fail) using taxpayers' money. In Europe, however, the crisis was grafted on to a previous situation characterised by the progressive loss of the role of the national financial institutions. The Community member-states cannot print money and therefore cannot use inflation to reward their exports and support development. They have to have resort to the dual instrument of issuing bonds and borrowing from other countries and from banks, also to fill the public deficit. Obviously, the sum received has to be returned with interest by a due date. This way the foundations have been laid to create what is usually called a two-speed Europe, divided between the creditors (Germany in particular) and the debtors.

The debtor countries which borrow money have to pay "premiums" which reflect the risk of bankruptcy (default), and the financial market even induces these countries to default through speculation (this is a simplified version and, for example, the risk of default for countries like Italy could have been much more serious in the absence of the euro, but I prefer not to venture on to the path of this hypothesis). There are naturally mechanisms of containment and constraint; the Maastricht agreements fixed the upper limit for the debt/GDP ratio at 60%, laying down fines for the countries that do not undertake rebalancing policies.

Policies of containment—austerity, in other words—were chosen to meet the very high public debt, leading to a lasting economic depression. In 1982, something similar happened when there was a serious crisis of banks, and the International Monetary Fund saved the banks by lending just enough money to the heaviest indebted countries to let them avoid default but at the cost of driving them towards long-lasting depression. Latin America, in particular, suffered from economic depression for a decade.

In short, the international economy now heavily conditions the policy of states. However, the vicious circle Europe is now in does not depend only on the conditionings due to globalisation of the economy and competition from the emerging countries but also on the imbalance between politics and finance. The cyclical fluctuations of the world economy were, until a few decades ago, limited by a protection system such as the Bretton Woods Agreement, part of the farsightedness and efforts of Keynes, which aimed at avoiding the repetition of world crises such as that of 1929. Keynes was not particularly satisfied by the outcome of the Bretton Woods Conference nor by the role played by the International Monetary Fund in the following years, but it is a fact that thanks to that agreement (even with all its limits), the developed economies enjoyed a period of stability, contained inflation and sustained growth.

Protection from cyclical crises had been made possible by the international rules being in harmony with the capacity of response by the states. Today, it is not only finance that is globalised, but so are the production of consumer goods, lifestyles and environmental and health risks. It would therefore be opportune to create an effective system of global protection, which energetically deals with the problems emerging in all these sectors, and not dismantling it, as has been done with Bretton Woods. (Obviously, it is out of the question to return to Bretton Woods, which reflected the political balances of the period.)

We run the risk—as with the end of Bretton Woods and the financial institutions inspired by Keynesian theories—of soon having to cope with the progressive erosion of supranational institutions such as the World Health Organization. The WHO has played a very important role since the end of World War II, establishing general guidelines and launching campaigns (like those for vaccination), which found in the different countries a network of public institutions at different levels which allowed them to be implemented. Today, the risk is that the initiatives of the WHO and other supranational organisations are neutralised by the lack of national public institutions of equal solidity, which can coordinate and make the preventive measures effective. There are signs in this direction not only in Greece, where the scaling down of the health system has affected first of all prevention, but above all in much of Africa.

The weakness of politics, in the face of the economy and the structural reforms it requires, cannot fail to have consequences on the weak attempts to stem climate change or prevent chronic diseases. It is worth remembering that already now for every dollar spent by the WHO on the prevention of diseases caused by food, more than $500 are spent by the food industry to advertise the products that foster them (www.foodcomm.org.uk).

An aspect that also needs to be mentioned is endemic corruption that affects the financial system and national and subnational governments, particularly in the health sector. Understanding of the financial crises has to be linked to the ways in which financial systems create structures that are inherently unstable and corrupt (with extremely high profit incentives for the financial system and risk borne by the general public). But corruption is also widespread, particularly in weaker countries, throughout governments and Ministries and affects political choices on health.

The WHO, Smallpox and Polio

Something similar to the crisis of the sovereignty of states may also take place in the health field. There are analogies between the economic crisis and what has taken place in some sectors of public health and above all what may happen in the future if supranational entities are dismantled or if they are no longer able to intervene energetically. By way of a concrete example, the Framework Convention on Tobacco Control, an agreement aiming to limit the trade and consumption of cigarettes, has been strongly opposed by the WTO. As it says on the website of the organisation: "In total, 45 specific trade concerns were raised, including 16 new ones, at a meeting of the Technical Barriers to Trade Committee on 24–25 March 2011. Tobacco regulations to protect public health continued to be a central concern. While members were not challenging public health objectives, they argued that the design of such regulations could have an unnecessary negative impact on trade." All that this legalese says is that freedom of trade has priority over the protection of health. Beyond the position of the WTO, almost the whole of Africa is disarmed in the face of the penetration of the tobacco industry (and soon of the food industry), a cause of substantial weakness both of its national institutions and the supranational ones.

Another example is that of vaccinations. In 1979, the WHO declared that smallpox had been defeated. In the early 1950s, millions and millions of new cases were registered every year and, even in 1967, 2 million people died from the disease. To stop the epidemic, a system of surveillance was started (which represented the greatest investment and the greatest success of the WHO) that allowed halting every new outbreak with isolating the cases and vaccinating the people who had come into contact with them. This goal was reached thanks to meticulous work done by the WHO in collaboration with national governments, and thanks to the fact that, from the early 1970s, the vaccine began to be produced in poor countries.

The case of polio suggests how the intensification of conflict and the relative loss of prestige of the international institutions are creating problems that are of a strategic nature. The campaign to wipe out polio by 2000 was launched by the WHO 25 years ago, and, although the objective was not reached, in these years extraordinary progress has been made: from about 350,000 cases in 125 countries in 1990, there were 416 cases in 2013, most of which were concentrated in three countries (Nigeria, Pakistan and Afghanistan), 359 in 2014 and less than 100 in 2015. In 2016 Nigeria celebrates 2 years without a case of wild poliovirus.

However, several facts should be noted. First, the latest successes—leading to potential eradication in the next years—are largely due to private intervention, an alliance between the Gates Foundation and funding from a number of countries (http://www.project-syndicate.org/commentary/polio-eradication-heroes-by-bill-gates-2015-12Polio; http://www.polioeradication.org/Financ ing.aspx). Second, failure to eradicate polio from the last remaining strongholds could still result in a large number of potential new cases. It is not a coincidence that the endemic countries are now Pakistan and Afghanistan, because they are at the centre of great military and cultural conflicts. In Pakistan, the refusal of vaccination has reached considerable dimensions, up to a maximum of 33% in the Swat valley, and this is linked to a series of prejudices spread by the Islamic fundamentalist forces: the idea that the vaccine contains oral contraceptives or derivatives of pork and more in general is an arm of the anti-Islamic forces to weaken the population. The same arguments were used in 2003 in northern Nigeria, where the failure of the vaccination campaign led to a new epidemic (Wilder-Smith and Tambyah 2007; Pallansch and Sandhu 2006). Third, the most recent and obscure case of an apparent fresh outbreak of polio—and the powerlessness of the international institutions—is that of Syria, where not only thousands of children have been killed in the current conflict, but many others have died from the return of diseases that could be easily avoided. Tuberculosis, diphtheria and whooping cough are increasing and polio has become an emergency in 2013. Polio had been defeated in Syria at the end of the 1990s after the compulsory vaccination had been introduced in 1964. There are no exact figures, but about 20 cases of poliomyelitis are believed to have arisen in 2013 (more according to other sources).

I am not focusing on AIDS here, but I just note that according to a recent report from the Kaiser Family Foundation and UNAIDS funding from donor governments fell in 2015 from $8.6 billion to $7.5 billion in 2014. This in spite of the fact that every year, around the world, nearly 2 million people (60% of them girls and young women) become newly infected with the virus (http://kff.org/global-health-policy/fact-sheet/the-global-hivaids-epidemic/).

In the last century, both the eradication of smallpox and the first phases of the campaign against polio were "apolitical", i.e. they were above political and military sides; the campaign against smallpox jointly involved the Soviet Union and the United States, with a substantial contribution by the WHO.

Globalisation and Health: The Prospects

Is it possible to establish a parallel between economic phenomena and health phenomena? There are certainly many parallels, especially if we examine the trends in time in different parts of the world. If for example we consider the growth in life expectancy in relation to income per capita, we can observe a sequence of phenomena which coincide approximately with economic trends: (a) from 1800 to 2011, there was a progressive increase in the per capita income and, in parallel, in life expectancy, but in a highly diversified way and with a growing gap, in particular between Africa and the rest of the world; (b) from 1950 onwards, there was a great acceleration in life expectancy in Europe and North America, which then experienced (in the 1970s and 1980s) a slowdown; (c) in 1950, China and India were still aligned with Africa for both life expectancy and income; since 1977 in both countries, there has been a growth in life expectancy totally divorced from income, whilst since 1985 the two indicators have begun to grow together.[1] These phenomena can be observed in the interactive graphic of Gapminder at the address shown in footnote 1. In particular, the reader can stop the clock in the four critical phases of 1900, 1950, 1977 and 1985 and lastly study the evolution up to 2012, to see the changes described here.

To what extent has economic globalisation contributed to health changes? Globalisation has been described as "a process of greater integration within the world economy through movements of goods and services, capital, technology and (to a lesser extent) labour, which lead increasingly to economic decisions being influenced by global conditions" (Jenkins 2004). It is a common opinion that this type of globalisation dates back to the early 1970s (the time of the first oil crisis) and is the consequence not of spontaneous tendencies but rather of conscious choices by economic and political elites (Labonté et al. 2009; see also Gallino 2013) and of the World Bank.

Has globalisation improved and can it improve the health of the world's population? If it is true that health depends to a great extent on the wealth of countries, globalisation ought to contribute to economic growth, to at least a partial reduction of poverty and to the release of resources and productive factors in the sectors of health and education.

Let us discuss first of all the initial assumption, i.e. that globalisation increases wealth. The value of world trade between 1960 and 2003 increased

---

[1] See www.gapminder.org. The website includes a great deal of interesting material. In particular on health v. per capita income. It can be accessed directly at http://tinyurl.com/ogrs3dp

from 24 to 48% of the global GDP. According to the usual interpretations, especially in the decade between 1980 and 1990, the economy of the countries that opened up to globalisation grew more quickly than in others. However, when looking more closely, it can be seen that growth of the countries taken as an example (China, Thailand, India, Vietnam) began thanks to closed and protectionist economies.

The theory of one of the greatest contemporary experts of globalisation, Dani Rodrik (2011), is that, contrary to the usual version, the countries with the fastest growth took off thanks to customs duties and other mechanisms protecting their industries, including state subsidies. The same theory is also upheld by a well-known economist who teaches at Cambridge, Ha-Joon Chang (2010): the countries that precociously opened up to globalisation and the ideology of the free market, i.e. Latin America and sub-Saharan Africa, grew much less than the protectionist ones (but whether there is a causal relationship is not very clear). In the 1960s and 1970s, Latin America grew at an average rate of 3.1%, whereas in the period 1980–2009 it grew by only 1.1.

According to both economists, the proportionality between growth and globalisation, understood as the degree of openness to world markets, is above all a construction of economists: if we exclude India and China (which have populations of extraordinary dimensions) in the period 1980–2000, the "globalisers" grew more slowly than the "non-globalisers".

The evidence is not at all clear either on the front of the reduction of poverty. Between 1981 and 2004, the value of the world economy was multiplied by four, and the number of people who lived with $1.25 a day or less was reduced by 900 million (from 1.9 billion to 1 billion). Can this be deemed a success? Not exactly, because this reduction took place almost entirely in China (where between 1981 and 2011 753 million people crossed the threshold of $1.25 per day), whereas the number of the poor did not decrease in sub-Saharan Africa. In addition, considering the threshold of $2 per day, the total number of the poor has remained almost stationary: 2.6 billion in 1981 and 2.2 in 2011 (source: World Bank, http://data.worldbank.org/topic/poverty).

Even taking for granted that globalisation fosters the overall production of wealth, growth has not been translated by redistribution, and conversely, globalisation has had concrete disadvantages. One common denominator of many policies in favour of globalisation and liberalisation (aka the Washington consensus) is containment of social expenditure and the promotion of private consumptions (in health as well): what is known as the commodification of essential needs.

The emphasis placed on the limitation of public expenditure, on levelling out the public debt, on the full recovery of costs through a contribution to expenditure by the users, etc., entails an implicit or explicit negation of the objective of keeping the welfare systems established after World War II alive, including universal health services. This is accompanied by a shift of emphasis from collective responsibility (and the promotion of health for all) to individual responsibility. The Declaration of Alma Ata in 1978 had sanctioned the universal right to the protection and promotion of health and well-being, but in 2013 the reform of Britain's historic and glorious NHS (National Health Service) by the Conservative government has led to the elimination from its charter of the sentence according to which the Minister for Health is bound to provide health care to citizens through the public service: today in Great Britain only individuals are responsible for their health (Davis and Tallis 2013). In 2013–2014, the NHS reduced its personnel, and this will also have repercussions on prevention.

There are other mechanisms through which globalisation can have negative consequences, especially for the poorest. Diderichsen et al. (2001) identified several main mechanisms through which globalisation generates inequalities in health: a growing social stratification and growing differentials in exposure to risk factors, in the likelihood and in the consequences of illness. In other words, globalisation expands the stratification between those who can benefit from it and those who cannot and at the same time creates social differences in exposure to risk factors—from smoking to junk food—and increases the likelihood for diseases through repeated and cumulative damage. Lastly, the descending trajectory expands at the individual level due to the reduced working capacity, the difficulties in obtaining insurance, the increase in insurance premiums and so on.

One of the main problems associated with economic changes is that the positive effects appear mainly in the long term, whilst the negative ones appear in the short term, often with catastrophic spirals for the most vulnerable individuals. The great changes that in the nineteenth century had led to an improvement in the hygienic conditions in English and German cities produced benefits for health more than half a century later, whereas the negative effects of the economic crisis for health are often immediate: the case of Nauru that we described is emblematic.

7

Cancer: A Time Bomb in Poor Countries

Cancer has been selected amongst several non-communicable diseases (NCDs) that are spreading in the world because it is representative of general problems encountered when studying NCDs and developing effective policies. Cancer has been extensively investigated (more than, for example, neurological diseases) and has several risk factors in common with other NCDs, so that many of the statements in this chapter could be generalised.

This chapter has a number of aims: to show that cancer is becoming a global problem, with a growing case load in low-income countries; to describe the main agents that cause it, and what is being done or what is not being done to reduce exposure to carcinogenic agents; and lastly to argue that primary prevention is by far the most effective and economic solution to fight cancer on a global scale. These considerations are also made in the light of political choices, such as the recent positions of the United Nations on the control of non-communicable diseases.[1] Further discussion on the strategic choices concerning prevention of tumours can be found in the article we published with Bray et al. (2015).

Every year in the world (around 2015), there are more than 14 million new cases of cancer, which causes 8 million deaths. At least 25 million people have had a diagnosis of cancer in the past 5 years and are still alive, and at least half of these people live in low-income countries. In most of these countries, a deep epidemiological change is taking place due to the joint presence of the traditional causes of death (in particular infectious) and the new degenerative

---

[1] "High-level Meeting of the United Nations General Assembly", September 2011.

© Springer International Publishing AG 2017
P. Vineis, *Health Without Borders*, DOI 10.1007/978-3-319-52446-7_7

pathologies typical (so far) of the advanced countries. However, the distribution of different types of tumour shows clear differences: whereas in the first positions in high-income countries there are cancers of the breast, of the colon and of the lungs, the low-income countries present much higher mortality rates for cancers of the stomach, liver and cervix uteri, all of which are widely or fully (cervix uteri) of infectious origin.

The frequency of tumours and other diseases varies according to geographical area and changes over time. For example melanoma (skin cancer) has a frequency that is 200 times greater in Queensland (Australia) than in Africa and China. In the 1960s, colon cancer was very rare in Japan, where today it records amongst the highest rates in the world.

This variability over the course of time and between geographical areas debunks a common prejudice that derives from a misunderstanding, i.e. that cancer is a genetic disease. The misunderstanding lies in the fact that the acquired genetic damage (e.g. mutations of environmental origin) is confused with changes transmitted by inheritance. In short, genetic is confused with hereditary. Tumours are unquestionably due to alterations acquired by the genetic material (DNA) but only to a small extent (from 5 to 10%) are they hereditary tumours. The alterations acquired but not hereditary can be mutations, gross chromosomal damage or functional changes of DNA without alterations of the sequence (we will look into the problem of epigenetic damage in depth in Chap. 8).

One piece of counter-evidence of the fact that most tumours are "environmental" and not hereditary comes from the study of migrant populations, because after a certain period of time those who emigrate acquire the same risk of cancer and other diseases of the area where they have settled. The case of Japan is well known: in the 1970s, stomach cancer was very frequent whereas that of the colon was not, contrary to what was observed with North America; however, when the Japanese emigrated to the United States, their rates of cancer rapidly fell into line with those of the local population, with an increase in tumours of the colon and a reduction in those of the stomach. Similar trends have been observed for many other degenerative diseases.

This is why the evolution of the state of health of the world as a consequence of social and economic changes must worry us, especially today when these changes have become so rapid and sometimes so unpredictable with globalisation. Before the spread of cigarettes, lung cancer was a rare disease and the consumption of cigarettes has spread relatively slowly, if compared, for example, with the changes in diet today under way in Asia or in Latin America.

Cancer as a Global and Preventable Problem

In this chapter, I refer to a classification of countries based on the Human Development Index (HDI), divided into the categories of low, medium, high and very high.[2] Although in the countries with a low HDI infectious diseases (including from unsafe sex) and diseases from malnutrition are still important causes of death, the projections suggest that by 2030 these diseases will have been overtaken by the so-called non-communicable diseases, including cancer (Bray et al. 2012; in actual fact, this definition is not valid for all tumours, as we will see). The estimated global load of malignant tumours will rise by 14 million new cases in 2012 to more than 21 million in 2030. The increase reflects the growth of the population and its ageing together with some important changes in the exposure to risk factors such as the increase of obesity and smoking.

There are great differences in the distribution of tumours in the world. If we consider the new cases diagnosed each year (i.e. the rates of incidence), there is a highly marked geographic variation depending on the income of the countries. In high- or very high-income countries, in 2008 breast, colon and prostate cancers represented 49% of the total, whereas in low-income countries tumours of the cervix uteri, of the liver, of the oesophagus and of the stomach were more frequent than in richer countries.

The rates of incidence of a specific type of tumour can also vary more than one hundred times between one geographical region and another. In the comparison between groups of countries, breast cancer has an average rate of incidence of 66 women out of 100,000 per year in high-income countries but drops to 27 in low-income countries (after having corrected for the different distribution by age of the populations), whereas cancer of the cervix uteri, the second commonest tumour in women at a worldwide level, is an example of the opposing distribution, i.e. of a greater incidence in low-income countries.

The rates of mortality also differ by geographical area, but the differences are less pronounced than those for incidence. For example, the rates of mortality for breast cancers are 15.5 out of 100,000 women per year in high-income countries and 10.8 in low-income countries, i.e. the incidence varies 2.4 times (66 against 27), whereas the mortality varies only 1.4 times (15.5 against 10.8), all other conditions being equal and after having corrected for the distribution by age. The data for cancer of the prostate are similar, with rates

[2] A longer and more detailed version of this chapter was published by the author in "The Lancet" (Vineis and Wild 2014), to which reference should be made for a more extensive bibliography.

of incidence equal to 63 in the rich countries and 12 in the poor countries (almost five times less) but mortality rates of 10.6 and 5.6, respectively (i.e. a ratio of less than twice). This distribution, which appears for almost all types of tumours, means that in the rich countries—compared to poor ones—for each person who dies of cancer there are far more who are given a new diagnosis.

This gap between incidence and mortality is due in part to the lower survival rates in the low-income countries. Africa, in particular, is very behind in terms of capacity for treating tumours: 15 African countries are completely lacking in pathology and radiotherapy services (Farmer et al. 2010; Sylla and Wild 2012) and in Gambia survival at 5 years for breast cancer is only 12% against 80% in rich countries and in some urban areas of China. However, the gap between incidence and mortality cannot be attributed only to less effective treatments in the poorer countries, but also to earlier diagnosis of the disease in rich countries, in part through screenings.

The Agents that Cause Cancer and the Importance of Prevention

The WHO report on non-communicable diseases—like the 25 × 25 strategy of the United Nations (see below)—lists amongst the main "behavioural" risk factors for cancer tobacco, alcohol, little physical activity and an imbalanced diet. However, cancer is a group of diseases that are far more heterogeneous than other non-communicable diseases and requires more structured and locally specific policies than those proposed by the WHO, which are focused on a small number of risk factors more relevant to the Western countries (Wild 2012).

Even though access to effective treatment and the development of new therapies are fundamental components in the fight against cancer, they will never be effective in the absence of prevention (Bray et al. 2015). A strong argument in favour of prevention is that thanks to it the causes can be removed permanently and the effort does not have to be renewed at each generation, as is the case for treatment. The role of prevention is particularly important where the resources are scarce, because it entails a substantial saving compared to the costs of diagnosis and therapies.

Another important property of prevention—which I would define "political" in the broad sense—is that for each subject involved in the preventive activity, many others benefit from it. The classic example is herd immunity: by vaccinating one person against an infectious disease, many more cases are prevented because the number of people who are infected and contagious is reduced with a domino effect. If the mechanism is clear for infectious diseases, a similar phenomenon can take place with the prevention of non-infectious diseases: the prohibition of smoking in public places not only exposes non-smokers less to passive smoke, but also induces smokers themselves to smoke less. In the same way, if an occupational carcinogen is abolished, it is not only the workers exposed at present who benefit, but also those who would have been exposed in the future, and even consumers.

The Global Epidemic of Diseases Linked to Tobacco

Tobacco is a powerful carcinogen with multiple effects and has a strong and growing impact on a planetary scale; it causes cancer of the lungs, of the upper respiratory tract, of the pancreas, of the stomach, of the liver, of the urinary tract, of the kidneys, of the cervix uteri and some forms of myeloid leukaemia. For some cancers (lung and upper respiratory tract), tobacco is the main cause, whilst for others it enters into a multifactorial mechanism together with other agents (like HPV for cervix uteri). It is also responsible for an important percentage of heart attacks and respiratory diseases, such as chronic bronchitis and emphysema. Tobacco is carcinogenic in more than one human organ, whether consumed in the form of cigarettes or in other forms such as bidis (the thin cigarettes smoked mainly in India), pipes or cigars. As well as finding a strong coherence between the observations in humans and biological evidence in other organisms or in cultured cells, today we know in detail many of the mechanisms of action of the carcinogenic substances in smoke, for example the formation of bonds with the DNA by the polycyclic aromatic hydrocarbons or the typical spectrum of mutations in key genes (such as TP53) which has been found in the tumours of smokers.

Smoking is currently responsible for about 30% of all the tumours in developed countries and of a lower, but rapidly growing, percentage in low-income countries. It is also responsible—and this is perhaps less well known—for a higher number of early deaths from cardiovascular and respiratory diseases than from cancer.

The fight against tobacco has been effective in some countries, to the point of making a substantial contribution to the decline of mortality. In the United States, between 1990 and 2009, the rates of mortality due to cancer decreased by 24% in men and 16% in women (Siegel et al. 2013; Eheman et al. 2012).

This improvement, which (to a lesser extent) is appearing in almost all the high-income countries, is due to early diagnosis, to slow but significant improvements in treatments and, to a great extent, to prevention: a great part of the decrease in mortality is concentrated in the tumours linked to tobacco. In particular, about 40% of the reduction of deaths from lung cancer in men in the United States between 1991 and 2003 was attributed to the reduced consumption of cigarettes in the past half century (Cokkinides et al. 2009).

Opposite trends, however, are observed in poor countries, where the absolute number of cases and the rates of incidence and mortality are increasing and are destined to further increase for many years to come, in part due to tobacco. The projections to 2030 of deaths caused by smoking tell us that all the increase will take place in low-income countries, where consumption is growing at worrying rates. If the current trends in the growth of consumption of cigarettes do not change, in the twenty-first century there will be more than 1 billion early deaths caused by smoking.

An action of containing tobacco sales and advertising in the poor- and medium-income countries is therefore absolutely essential as from now. Whilst the full implementation of the Framework Convention on Tobacco Control by the countries that have signed it is vital for successful prevention, unfortunately it is hindered by the WTO in the name of free trade.

Occupational Carcinogens

From 1921, when the International Labour Office (ILO) introduced the first specific regulations, great progress has been made in the prevention of occupational tumours in the high-income countries. In these countries, the elimination or a substantial reduction of exposure to asbestos, aromatic amines, benzene, benzidine and other carcinogens has prevented tens of thousands of cases of cancer. For example, there is convincing evidence of the fact that a decline in bladder cancers and leukaemia amongst workers in the United States and Great Britain followed on respectively to the banning of aromatic amines and benzene.

The effects of exposure to asbestos persist for decades after the end of exposure, and the peak of mesotheliomas and lung cancers that can be attributed to asbestos in many countries has not yet been fully observed. However, a decline in the risk of mesothelioma has already been described in the United States and in Sweden, where restrictive measures for the control of asbestos in workplaces were introduced in the early 1970s. Resistance to change, however, is strong and strange arguments (to say the least) are heard, like the fact that Indians would not be susceptible to asbestos, which is obviously wrong (Krishnan and Ray 2010).

Despite the progress made, occupational cancers remain a political priority, considering the unequal distribution of risk, which affects only some categories of people. One of the most urgent problems to be faced is the fact that productions using carcinogenic substances now tend to be exported to low-income countries, such as asbestos by Canada and various toxic substances by Japan (Kirby 2010; Park et al. 2009).

Diet and Tumours

Obesity is an important risk factor for breast, colon, endometrial, kidney, oesophagus and pancreas cancer (Cogliano et al. 2011). Alcohol is clearly associated with cancer of the liver, of the upper respiratory tract, of the breast and of the colon (WCRF/AICR 2007). A diet poor in fibre has been related to tumours of the colon-rectum (WCRF/AICR 2007). However, despite the progress and a massive body of research done in the past decades, the comprehension of the relations between diet and tumours is still fairly uncertain, in particular as far as the mechanisms of action and the role of the individual nutrients are concerned.

Recommendations based on a systematic review of evidence have been published by the World Cancer Research Fund (WCRF/AICR 2007) and suggest that regular physical activity; a reduced consumption of fats, sweetened drinks, alcohol and salty foods (salt is a risk factor for tumours of the stomach as well as the main cause of hypertension); a varied diet rich in fruit, vegetables and pulses; and keeping weight in the norm can prevent a number of tumours that is by no means negligible. All these modifiable risk factors reduce risk of several cardiovascular diseases (CVDs), and this is where the "disease by disease" approach fails—the benefits for cancers, CVD and chronic respiratory diseases make the cost-effectiveness ratios much more advantageous than for any single disease.

The EPIC (European Prospective Investigation into Cancer and Nutrition) population study, in which more than 500,000 Europeans were studied, has showed that if the recommendations of the WCRF were regularly followed, the risk of tumours, in particular of the stomach and colon, could be significantly reduced (Romaguera et al. 2012).

Infectious Carcinogenic Agents

According to a recent analysis (de Martel et al. 2012), at worldwide level the proportion of tumours attributable to infectious agents is of 16%. However, this fraction is much higher in low-income countries (23%) than in high-income ones (7%) and varies from 3% in Australia to 33% in sub-Saharan Africa. *Helicobacter pylori*, hepatitis B and C viruses and the papillomavirus (HPV) are responsible for an important percentage of tumours of the stomach, liver and cervix uteri, respectively. HPV is very probably responsible for all the tumours of the cervix uteri, i.e. it is a "necessary" cause of it. About 30% of tumours of an infectious nature occur in people under 50.

These figures alone show the need for a more articulated approach than that adopted by the United Nations and the WHO to chronic diseases, based exclusively on behavioural factors, considering that part of them in actual fact have a communicable cause and affect poor countries to a great extent.

One of the most important achievements in recent years in cancer research has been the development of an anti-HPV vaccine to prevent tumours of the cervix uteri. Ensuring that the vaccine is available to the poorest populations, who are those with the highest incidence of tumours of the cervix uteri, is decisive. The problem is not marginal, because the recommended coverage by the vaccine is greater than 70%, but the cost is still very high. An accurate evaluation of how mass vaccination will interact with screening activities (Pap test or test of the DNA), for example in adolescent girls, is also necessary.

Another effective vaccine is that against hepatitis B, and in this case too, it is particularly important that it becomes available in poor countries where the frequency of infection is higher. A reduction in the incidence of liver cancer has already been observed in the countries that introduced the vaccine in the 1980s.

Environmental Carcinogens

The extent of exposure to environmental carcinogens in the world is unknown, in particular in low-income countries, although the total figure of tumours that can be attributed to them may number hundreds of thousands, and only limiting the calculation to known carcinogens for which data on exposure are available (arsenic, atmospheric pollutants, aflatoxins, polychlorinated biphenyls—PCB, asbestos). The effects of additional exposures, such as to heavy metals (chromium, cadmium, nickel, beryllium) and to other carcinogens are difficult to quantify because we do not have adequate information on the spread and number of people exposed in low-income countries (Vineis and Xun 2009).

Another subject which is still controversial is that of exposure to electro-magnetic fields in relation, for example, to the use of cell phones. The working group of the IARC Monographs which carried out the critical analysis of the studies judged them as a whole "limited" for brain tumours (gliomas and neuromas of the acoustic nerve) and "inadequate" for other types of tumours (www.iarc.fr).

Environmental exposure is an area of great potential impact on public health, but for many kinds of exposure the scientific evidence is still conflicting or uncertain. In spite of massive public attention in the media, studies of environmental exposures are very difficult to conduct and the findings are often not robust despite the best efforts of scientists. This is due to several reasons: information on individual exposure is often poor or cannot be ascertained easily (e.g. to estimate individual exposure to PCBs—an almost ubiquitous contaminant—we need measurements in blood or in fat); cancer arises after decades since exposure starts, so that we need longitudinal studies that last decades; single types of cancers are usually rare, so that very large populations such as EPIC should be investigated; and exposure usually occurs at low levels, meaning that we expect a modest increase in risk. Causal assessment encounters difficulties such as those described in the chapter on climate change, including the lack of experimental studies for ethical reasons.

Atmospheric pollution is one case where the quantitative and qualitative increase of research in recent years (in particular thanks to the ESCAPE multicentric study on the chronic health effects of pollution) has allowed dissipating many uncertainties, showing that air pollution—due mainly to traffic—has numerous and important repercussions on health, in particular for cardiovascular diseases, for the respiratory system and for lung cancer (Beelen et al. 2014a, b; Raaschou-Nielsen et al. 2013). All the more so, the disease load linked to air pollution is massive in countries such as China where levels of particulate more than five times greater than in Western countries are often recorded.

Other Risk Factors: Socio-economic Status

In high-income countries, tumours clearly show a higher incidence and mortality in the lower socio-economic groups, and this distribution cannot be explained entirely by the known risk factors, indicating that something else (still unknown) underlies the effects of social stratification (for this, see the website http://www.lifepathproject.eu and Vineis et al. 2014). For example, in the EPIC study, we observed that lower education is associated with a higher mortality rate for all causes (Gallo et al. 2012) and a greater incidence of cancers of the stomach, of the oesophagus and of the lungs (after having deducted the effect of smoking). Risk factors such as obesity or lifestyle are also associated with the socio-economic status.

"Why Me?" Nature and Culture: The Role of Genetic Predisposition

The season of systematic research into the role of genes in the onset of diseases has not yet produced encouraging results about the possibility of "personalising" the prevention of cancer as it was hoped some years ago. Although important discoveries have been made on the role of genetic predisposition, these are significant especially for the study of the molecular mechanisms of the onset of tumours, rather than for applications in public health. One example is the discovery of variants in the 8q24 region of the chromosome 8, an area defined "gene desert" as it does not contain genes but sequences of DNA that are perhaps more important than genes. This region is involved in cancers of the prostate, breast and colon but also in diabetes and cardiovascular diseases. The studies on 8q24 are very interesting and may lead to new and extraordinary discoveries but for the time being have few practical consequences on the prevention of tumours.

One of the misunderstandings that underlie the nature/culture dichotomy (genes/environment) is the lack of understanding of the fundamental difference between variation and causality. Dr. Geoffrey Rose pointed out years ago that if everybody smoked a pack of cigarettes per day then lung cancer would be considered a genetic disease, because the role of tobacco as the main risk factor would not be identifiable, whereas the role of genetic causes alone would emerge to explain individual variability. In actual fact, the last decade of systematic genetic studies has not confirmed the hypothesis that genetic differences play an important role in lung cancer: if today we were to eliminate smoking, the main cause of lung cancer at the population level would be atmospheric pollution, not genetics.

Today there is a vast amount of literature on the fact that some diseases occur more frequently within some families and can therefore have an important hereditary component. However, only a small part of diseases has been explained by the recent, impressive investment in what are known as genome-wide association studies (GWAS), i.e. systematic studies of association between the genome and diseases. This investment came from the observation that many diseases recur more frequently in families. One of the main reasons for the failure of GWAS to explain the recurrence in families is that we inherit from our parents not only the genes but also the environment in which we grow up and the culture that we receive: for example, the children of smokers are more often smokers themselves, and even the preference for certain foods is transmitted through example and education. Research on the "missing heritability" (the difficulty of finding genetic differences which explain the frequency of tumours in families) and on "nature v. culture" seeks, in essence, to separate two categories of causes (the genes and the environment) which are closely connected and cannot easily be disentangled.

Primary Prevention of Tumours

In practice, only a proportion of between 5 and 10% of tumours is due to mutations transmitted from parents to their children, in high-risk families. Typical examples are breast cancers transmitted by the BRCA1 gene, or a rare tumour of the retina in children linked to the gene Rb1 (retinoblastoma). Another type of genetic variant transmitted by inheritance, on the other hand, modulates the individual risk of cancer but without playing an autonomous role, in the absence of environmental exposure. These are defined "low-penetrance variants" (penetrance is the strength with which a gene variant can cause the disease). It is now evident—also according to the studies on migrant populations—that as a whole the environment plays the far larger part in causing non-communicable diseases.

For a long period after the publication in 1981 of the book by Doll and Peto *The causes of cancer,* which contained estimates for the United States, the percentage of tumours attributable to environmental risk factors, i.e. not genetic, was a very controversial subject. By environmental, we mean here exposure to external agents, including behavioural such as smoking and alcohol consumption, but also internal, for example, metabolic causes: in practice everything that does not include inherited genetic mutations or variants. The most recent and accurate estimates have been proposed by Max Parkin who, in a monograph on cancer in England, considered fourteen risk factors and eighteen types of cancer (Parkin et al. 2011). It appeared that 45% of tumours in men and 40% in women would be avoidable with reasonable preventive measures, the most drastic of which is the abolition of smoking. Such preventive measures would substantially also reduce the load of cardiovascular diseases, diabetes, obesity, liver and kidney diseases and probably neurological ones as well.

The Threats of Globalisation for the Primary Prevention of Cancer

In September 2011 at the "United Nations High-level Meeting on Noncommunicable Disease (NCD) Prevention and Control", the world's leaders committed themselves to energetically facing up to the threatening epidemic of degenerative chronic diseases. The main concerns include the great economic and social consequences of the epidemic. Eight months later, the Assembly of the WHO set the objective of reducing the mortality rates for NCD by 25% by 2025. Known as the 25×25 strategy, it has been incorporated into the WHO plan of action for 2013–2020, which in turn lists nine national objectives. Two objectives are general: to reduce mortality from NCD and halve the increase in obesity and diabetes. The other seven are specific: a reduction in the consumption of alcohol, salt in the diet and smoking, control arterial pressure, an increase in physical activity, greater access to pharmacological treatment of people at a high risk of cardiovascular diseases and a wider access to basic technologies and essential medicines. It is to be noted that at national level these objectives are "voluntary", i.e. the States have to find the resources to implement them.

As can be seen, the strategy is focused almost exclusively on "behavioural" risk factors typical of Western countries: tobacco, physical exercise, obesity, diet and excess salt. The identification of these priority risk factors is hard to disagree with, but there are serious limits in the 25×25 strategy (see the criticism by Pearce et al. 2014). There is evidence of the fact that prevention based on a strictly individualised approach—such as educational messages by healthcare personnel—has an overall modest impact on cancers, whereas actions at the level of the population are more effective. This is shown by the role played by the structural measures implemented by Mayor Bloomberg in New York, which we will discuss in the last chapter. Other examples of effective structural interventions are the reduction of smoking in Papua New Guinea through price increases and a substantial reduction in the levels of cholesterol in Mauritius thanks to trade agreements which have allowed shifting consumption from palm oil (rich in saturated fatty acids) to soy oil for cooking.

There are reasons for concern about the future of primary prevention because the tendency to reduce public expenditure and privatise part of the health systems will have consequences on the preventive activities, which are not remunerative from the economic point of view (possibly with the exclusion of early diagnosis) and therefore are not attractive for the private sector. The reduction of budgets for primary care will probably have a negative influence on the inclination of doctors to commit themselves to prevention. In the United States, Europe and Canada, less than 4% of public expenditure on health is for cancer prevention (in all its forms) (Sullivan et al. 2012). In addition, universities are encouraged to work with industry to develop new therapeutic treatments or new technologies, whilst few profits can be obtained with primary prevention.

If the approach based on the free market has spread risk factors and lifestyles of the Western countries throughout the world, including high levels of consumption of alcohol, cigarettes and junk food and little physical activity (with great differences between social groups), there is not a corresponding spread of globalised messages of prevention. Primary prevention tends for the time being to be diffused with antiquated methods. When drastic measures were introduced, as in New York by the Mayor Bloomberg, they were criticised in the name of individual freedom, as we will see in the last chapter.

Final Remarks

Cancer is a growing global health problem, but it does not have a uniform distribution. I have maintained that primary prevention is by far the most effective and least expensive means to drastically cut the global risks of cancer; between one-third and half of the tumours in the world could be avoided with primary prevention, a figure that is far higher than that of deaths avoided thanks to therapies. There are some great advantages in primary prevention: its effect extends beyond the people directly involved (the herd effect); the impact can be transgenerational (preventive measures—unlike therapies—do not have to be renewed with every generation): the elimination of environmental or occupational carcinogens can be definitive whilst the expenditure on drugs appears with every generation of patients; lastly, avoiding exposure to carcinogens allows preventing other chronic diseases. However, the current social and political climate does not play in favour of prevention. Global styles of consumption, supported by changes in the global economy (e.g. in the quality of food) and in behaviour (the low amount of physical activity associated with the built environment), are not contrasted by equally global initiatives. The emphasis on the privatisation of social services and health systems will probably have a negative effect on primary prevention and accentuate social inequalities.

8

The Epigenetic Landscape

This chapter deals with the lasting marks that the great transformations linked to globalisation can leave on our DNA. This is no metaphor: it is possible that epigenetic alterations will be produced by globalisation and, in the future, become one of the fields of greatest development of health studies. The theory of this chapter is that the great changes in the markets, in the availability of industrial food and in other important aspects of lifestyle which have taken place in the past 30–40 years might have an impact not so much on the genetic make-up but on what we call the "epigenetic landscape".

To explain what epigenetics is, it is worthwhile starting from genetics, i.e. from the hereditary transmission of phenotypic features through the DNA sequence. There are some clear examples of how the environmental changes of the past have had consequences on the genetic characteristics of certain populations, selecting gene variants. For example, the migrations from the Fertile Crescent of the Middle East to northern Europe, which took place between 5000 and 10,000 years ago, led to selection among the new settlers of the northern countries of the traits for tolerance to lactose and the diffusion of fair skin. Both mutations emerged in all likelihood to make up for the deficiency of Vitamin D, due to the reduced exposure to the sun in northern countries. A deficiency of Vitamin D drastically increases the frequency of rickets and consequently the risk of death during labour due to the pathological conformation of the pelvic bones. It is thought therefore that fair skin—which makes the ultraviolet rays more effective in activating provitamin D—and the capacity to digest milk in maturity were two mechanisms of adaptation which appeared through selection of favourable genetic mutations in the

© Springer International Publishing AG 2017
P. Vineis, *Health Without Borders*, DOI 10.1007/978-3-319-52446-7_8

migrant populations (even though doubts have recently been raised on the tolerance to lactose). Similarly, the heterozygotes for the gene G6PD or for sickle-cell anaemia, which are more frequent in areas traditionally affected by malaria, such as Sardinia once was, have been selected because the subjects carrying the mutation are less predisposed to the action of the Plasmodium that transmits malaria.

It is very probable, however, that it will not be the slow changes in the genetic make-up of the populations, i.e. in the DNA sequence, which will dominate the stage following the rapid changes linked to globalisation. It will be the faster epigenetic changes which are only now beginning to be understood in detail. These are functional changes—not structural, i.e. not in the sequence of the DNA bases—which are reversible and transmissible from one cell to its daughters. There is increasing evidence of the fact that the epigenetic changes of the DNA are linked to environmental exposure and in particular to exposure in the womb. The transgenerational transmission of epigenetic changes (not simply from cell to cell but from whole organisms to their offspring) is a highly controversial issue. Though there is evidence in plants and in certain animals, there is no evidence so far in humans, and the hypothesis of transgenerational changes should not become a new Lamarckism (Heard and Martienssen 2014).

The evolution of living organisms is the expression of different and at times opposing forces. The main tension is between the need for stability by the organisms and the need for change to adapt to changing environments. On the one hand therefore, we have the great structural stability of the DNA molecule and also the semantic stability of its code, preserved practically intact through the species. On the other hand, to adapt to a constantly changing environment and to the continuous threats that these changes potentially represent, mechanisms have been selected which guarantee sufficient variability of the genetic configurations, which in turn allow adaptation and evolution.

The protagonists of these mechanisms of flexibility are the transposable elements (transposons), the crossing-over in meiosis (cellular division of gametes) and epigenetic changes. The great merit of having understood the importance of the transposable elements and epigenetic research is due to Barbara McClintock (Nobel Prize in 1983), after the foundations for this research had been laid by Conrad Waddington. In 1942, Waddington defined epigenetics as the "the branch of biology which studies the causal interactions between genes and their products, which bring the phenotype into being". To explain the capacity of adaptation of maize to the hybridisations and the selections made by farmers, McClintock hypothesised the existence of "jumping genes", which today we call transposable elements. On the other hand, McClintock herself in 1951 had the intuition that the expression of DNA was allowed by mechanisms similar to those identified today in the methylation of DNA and in the acetylation of histones: "This inactivity or suppression is considered to occur because the genes are "covered" by other nongenic chromatin materials. Gene activity may be possible only when a physical change in this covering material allows the reactive components of the gene to be "exposed" and thus capable of functioning" (McClintock 1951). The anticipatory nature of these statements will be understood shortly.

As a general rule, repeated environmental "stresses" tend to increase the instability of the genome, for example through the demethylation of DNA or the growing activity of transposons, which on the other hand allows increasing the genetic diversity to better respond to the environmental stresses. This was how, according to McClintock, the maize reacted to the continuous "selective stress" induced by farmers. Genomic instability, however, is also a potentially harmful mechanism for the cell and for the organism and can foster an increase in diseases.

In the past two decades, many efforts have been dedicated to research on the epigenetic changes that accompany the development of diseases such as cancer. The most widely studied epigenetic mechanism is the methylation of DNA. Observed from the early 1980s in research on carcinogenesis, methylation of DNA consists of the addition of methyl groups to the bases of DNA and causes a repression or silencing of the activity of the genes, according to what Barbara McClintock advocated. If the silencing concerns the class of so-called oncosuppressor genes, it can lead to the development of cancer. Both hypermethylation of the oncosuppressor genes and the demethylation of the so-called oncogenes have the same effect. Widespread demethylation of the genome is also accompanied by a generalised instability of the genes and an increase in their expression (translation into proteins).

Demethylation and the consequent instability of the genome increase the gene variability and are therefore a double-edged mechanism, which on the one hand can allow expressing the genetic varieties useful for overcoming an environmental stress, but on the other can foster diseases. The mechanisms described here with reference to cancer are probably also at work in many other diseases.

Another epigenetic mechanism that is significant for the study of environmental damage—but which has been studied far less than methylation—is the modification of histones, proteins that cover the DNA and take part in regulating its expression. Not only has an alteration in the functionality of the histones been observed in certain forms of cancer, but some environmental exposures, for example to arsenic, seem to act above all through a modification of the histones. Experimental and epidemiological studies have shown that arsenic is an important pulmonary and bladder carcinogen and is—as we have seen—a contaminant of drinking water in vast areas of the world.

Lastly, mention must be made of the microRNA (or miRNA), short chains of RNA (around 22 bases): there are many of them (probably thousands), and they move along the genome (downstream from the transcription of the DNA in messenger RNA), to regulate their expression according to requirements. Once again, the importance of the miRNA is linked not only to their involvement in the onset of tumours but also to the modifications they undergo due to environmental exposure. The methylation of DNA is however the mechanism which has been most studied in relation to environmental exposure, including exposure to food.

Crucial epigenetic changes, due to the environment, can take place as early as in the uterus; "epigenetic programming" plays a crucial role in embryonic development and in conditioning the physiological development and the health of an individual in the long term. For example, a study on female rats exposed to arsenic during pregnancy showed that this element induced epigenetic alterations (patterns of methylation of the DNA) in the brain cells designated for the memory, which persisted in the course of the adult life of the offspring (Martinez et al. 2011).

Diabetes, Birth Weight and Epigenetics

Prospective observations on groups of infants show that malnutrition, a low birth weight and rapid postnatal growth are all phenomena associated with a high risk of diabetes in adulthood. The risk is further intensified if the children with a low birth weight have a (too) rich diet, of the Western type, during growth and in adulthood. The interpretation that is usually given is "foetal programming", i.e. the embryo of undernourished mothers is "programmed" according to an environment poor in nutrients and therefore tends to accumulate energy. This is the theory of the so-called thrifty phenotype: when, as a child or an adult, the same individual is exposed to an abundant or overabundant diet, the accumulation appears in the form of obesity, peripheral resistance to the effects of insulin and diabetes (diabesity). There is beginning to be concrete evidence of the fact that these pathological outcomes are mediated by epigenetic modifications.

Many of the cases of low birth weights are in low-income countries with the greatest incidence in South-East Asia (31% of births), where at the same time a transformation of the diet in a Western and industrial sense is under way. The results of a study on maternal nutrition in rural areas of Southern Asia have shown that mothers were shorter and thinner than European ones and that term babies were 700 g lighter than the European average. Investigations in the same areas now confirm the association between a low weight at birth and the development of an altered tolerance to glucose and type two diabetes.

According to the results of an Asian study, the risk of diabetes was six times greater in individuals whose BMI (body mass index) as children belonged to the lower third and who as adults moved into the upper third, compared to individuals with the opposite trajectory, i.e. who were overweight as children and lost weight as adults.

Diet and Epigenetics

Diet influences the state of methylation of DNA in different ways. One of the major subjects of research is the cycle defined 1-carbon metabolism, which has the main responsibility for the donation of the methyl groups at cellular level. A deficiency of folate, one of the key components in the cycle as well as a constituent of food, has been associated with an increased risk of cancer and other degenerative diseases, as well as congenital neural tube defects. One of the mechanisms through which a deficiency in folate can influence carcino-genesis is hypomethylation of DNA and therefore the instability of the genome. Alcohol, a risk factor now known for breast cancer, reduces the bioavailability of folate. Although these are hypotheses to be ascertained by further research, it is probable that excess alcohol and a deficiency in folate are connected mechanisms in the processes that lead to some types of cancer.

One eloquent example of the impact of epigenetics through diet has been observed in bees. The completely different castes, from the behavioural and reproductive points of view, of the queen bee and the worker bees are the result of a differentiated consumption of royal jelly by the larvae. Royal jelly influ-ences the methylation of DNA and the degree of methylation "directs" the development to one or the other caste. Some Australian researchers have recently imitated the effects of royal jelly by "switching off" the enzyme that transfers the methyl groups to the DNA in the larvae of bees. The larvae were transformed into queens, in the total absence of royal jelly, but exclusively due to an intervention on the enzyme (Lyko et al. 2010; Kucharski et al. 2008).

The effect of maternal malnutrition on human offspring has been studied amongst Dutch people whose mothers suffered hunger during the Second World War (Dutch famine of 1944). The children of the women who in pregnancy had suffered from the famine had a much higher probability of developing metabolic diseases (diabetes and obesity) compared to the children of women who became pregnant before or after the famine. Epigenetic analyses carried out on the children when they reached the age of 60 showed a differential methylation of various genes involved in growth and in metabolic control, which were also variable in relation to the gender of the infant and the trimester of the pregnancy in which the woman had suffered from the famine. A reduced methylation of IGF2, a key gene for the metabolism and for predisposition to diabetes, has been observed in individuals exposed to famine in the peri-conceptional period, with respect to siblings who were not exposed (de Rooij et al. 2007; Heijmans et al. 2008).

In another study, the individuals who were in the womb in the years of the famine which accompanied the civil war in Nigeria (1968–1970) 40 years later had an increased risk of hypertension, had an altered tolerance to glucose and were overweight. Similar observations have been made in relation to the famine of 1959–1961 in China.

Can Diet Stabilise the Genome?

The best way to establish a cause–effect relationship is to carry out an experiment, possibly randomised, based on the comparison between two groups—treated and controls—and on the random allocation of the treatment in the two groups. In one randomised study on diet, we wanted to test whether flavonoids, a class of compounds found in fruit, vegetables and pulses, had an epigenetic effect and could stabilise the genome. We studied the methylation of some key genes in carcinogenesis, such as RASSF1A and MLH1, and the repeated LINE-1 sequences, where the transposons (transposable elements) are to be found.

In our study, we observed an increase of the methylation in the peripheral cells (white blood cells) in association with the regular consumption of fruit and vegetables, although the effect did not seem attributable in particular to the flavonoids. We also observed a reduction in the intra-individual variability in repeated samples, in agreement with the hypothesis that consuming fruit and vegetables (and methyl groups) can effectively stabilise the genome (Scoccianti et al. 2011). The effect was particularly marked for LINE-1, i.e. for the global methylation and for the areas of the DNA that deliver the transposons. A reduction of the methylation in LINE-1 was associated by previous research with exposure to risk factors (such as atmospheric pollution, for example) for cancer and cardiovascular diseases.

These observations thus reinforce the hypothesis that environmental stress has a destabilising effect on the DNA, whilst some dietary behaviour would have a stabilising effect, i.e. a beneficial one. We have considered only diet, but similar effects have been observed in other studies on physical exercise (Zhang et al. 2011).

Globalisation and the Epigenetic Landscape

If we consider some important phenotypic changes that have taken place in the populations of the past, we can observe that many of them were linked to mass migrations: tolerance to lactose, the colour of the skin, diabetes in Asian migrants to England, melanomas in Queensland, tumours of the colon in Japanese migrants, etc. Migrations are one of the most visible characteristics of globalisation, but interest in migrations goes far beyond the short term. It seems in fact that the world's populations as we know them today are the fruit of long periods of adaptation to the environments in which they have lived, more or less stably. The phenotypes of the Pigmies, of the Watussi and of the Inuit express adaptations to local environmental conditions, probably as a consequence both of genetic selection and of epigenetic adaptations.

However, living conditions are now changing all over the world much faster than in the past (the populations "segregated" from civilisation are now an extremely small minority), and it is improbable that adaptations involve the appearance in short periods of time of genetic mutations that spread to a large share of the population. Population genetics takes a very long time, and it is probable that in response to globalisation (including migrations), the more rapid epigenetic mechanisms will be involved (Vineis et al. 2014). It must be noted that in the past it was predominantly individuals who moved towards new environments and therefore had to adapt; now it is also the environments that "go" to individuals through economic and commercial globalisation.

Two examples of rapid and marked changes which started in the West and are now being transferred to the rest of the world are the lowering of the age of the menarche (and therefore the fertile age) and the increase in average height. These changes are part of a wider picture of hormonal and sexual transformation, as illustrated by Gluckman and Hanson in the excellent book *Mismatch* (2006), which also highlights how increasingly early sexual and reproductive maturation is mismatched from psychosocial maturation. In other words, biological maturity is reached today much earlier than in the past, whereas psychosocial maturity—such as the sense of responsibility and the ability to maintain a family—has been, on the contrary, dramatically delayed. In the global economy, the market commercially exploits the "infantilisation" of society and early sexual maturation through a world of commodities addressed to the so-called pre-teens whilst it seems incapable of providing professional opportunities to young people until they are thirty or older.

The transformations that have been observed for some time in Western countries in the hormonal and reproductive characteristics are now clearly visible in China as well, for example, with a rapid anticipation of the menarche. The globalised hormonal and sexual changes are probably linked to other rapid transformations taking place in the world, of a cultural, behavioural and dietary type, and it is probable that they are mediated through an epigenetic mechanism, as suggested by some recent studies (Demetriou et al. 2013).

9

The Political Choices

"This is the idea that, despite the immense differences in history, culture, language and geography that divide them, there could develop in the citizens of the [European] Union the profound sense of belonging to a project of civilisation and of social progress of which there is nothing comparable in the world. It is a project that identifies with a notion of democracy as a political system in which all the members of the community have both the right and the material possibility to intervene effectively and in a participatory way in making the decisions that affect the production and distribution of public assets, such as those incorporated by the European social model, on which not only the materiality of their existence but also the very ultimate meaning that they would like to attribute to it, depends." (Gallino 2013, p. 225).

Globalisation has been analysed in particular from the economic and social points of view, whereas to date that of the health of the populations has been little studied, even though it is probable that it is in this very field that important effects will be felt. For example, it is possible that for the first time, the state of health, in particular in the high-income countries, locally gives signs of decline. And if there is no decline (which cannot be proven at present), it is likely that there will be greater instability in the "health system". Many countries are seeing an increase in the gap between social groups, though the overall image is complex depending on the metrics used (Atkinson 2015 and the Equality Trust, https://www.equalitytrust.org.uk/about-inequality/scale-and-trends). Another consequence of globalisation is the increased importance of private interests and pressure on public research institutions and on health

© Springer International Publishing AG 2017
P. Vineis, *Health Without Borders*, DOI 10.1007/978-3-319-52446-7_9

systems where business and politics are interlocked. This tendency at present appears destined to become even greater rather than contract. This trend is already very clear in England today.

The End of Solidarity?

There are various indirect ways through which the economic crisis and the social climate can influence health. Prevention has several advantages over therapies, the main one of which is the fact that its effects can last indefinitely; i.e. prevention does not have to be renewed at each generation. Banning an environmental or an occupational carcinogen is conclusive, whereas, without prevention, with each generation there are new patients who require new therapies. In addition, unlike therapies and genetic screenings, preventive actions are often effective for more than one disease: diet and physical exercise have a positive effect on several types of tumours, on cardiovascular diseases, on diabetes, on hypertension and probably also on neurological diseases. However, to be effective, prevention has to be based on actions at collective level, whilst the usefulness of the strictly individual approaches is generally limited to the higher social groups (to be clear: a healthy diet, physical exercise, consuming small amounts of alcohol and not smoking are habits absolutely to be pursued at individual level, but structural measures are necessary for them to take place in all the strata of the population).

Unfortunately, the current economic and political climate does not facilitate a commitment for prevention. The trend towards privatising health, as is taking place even in Great Britain, the home of the glorious National Health Service (NHS) (Davis and Tallis 2013), means that doctors will have less time and interest in promoting health. Already at present, the share of public health spending on cancer prevention (screenings included) in North America and in Europe is less than 4%. With the privatisation of health, prevention would not be very appetising because, save some exceptions, it does not generate private profits, even though guaranteeing enormous public savings.

On the moral level, the principles of solidarity that had taken root in Europe after the Second World War, or even before it (in particular with the mutual aid societies and social insurance), appear obsolete to many today. The division of countries into creditors and debtors sheds a negative light on the latter (in German "Schuld" means both "debt" and "fault"!) and is accompanied by a scaling down of their welfare systems. More in general, the crisis increases the drive for localism and fosters the emphasis on private consumption rather than on public services, motivated by the twofold objective of supporting industrial production and reducing public expenditure.

The new trends in the economy and in politics can undermine solidarity at its base even in subtle but more serious and lasting ways, because citizens may no longer understand why they should respect the law and pay taxes. What should we think of the extreme forms of injustice that are committed every day? For example, the fact that American taxpayers paid $6 billion to save Goldman Sachs, a company which the year afterwards spent at least $2.6 billion on bonuses for its top managers? The free market ideology and the associated utilitarian ethics have led to the erosion of some traditional values. The tendency of utilitarian approaches is to replace moral values as such (as interpreted for example in the Aristotelian or the Kantian tradition) with quantifiable entities, in the context of a presumed rational calculation of the marginal costs of certain actions with respect to their marginal benefits. For example, Gary Becker, 1992 Nobel Prize winner for Economics, has maintained a theory of marriage as a simple commercial transaction based on mutual convenience (costs to benefits ratios), coherently with other contractual-like tendencies of our society.

Concepts such as love and responsibility for others and the future genera- tions are at stake, with respect to which rational (or presumed so) calculation becomes predominant, according to economists like Becker. Even accepting gambling as an integral part of the world of finance, but also of everyday life— with the proliferation of betting centres imported from the English-speaking world— implies a conception of life in which a fortune can be accumulated quickly without investing in individual and social responsibility. As an Amer- ican manager has said, "If you find an executive who wants to take social responsibility, fire him. Now" (DeVogli 2012).

Coherently with this "moral climate", the conviction is widespread amongst some economists who are advocates of the free market that the increased mobility of capital has made welfare policies impracticable, in particular in Europe. The strength of speculative capital has increased enormously, and this hampers the ability of States to spend into welfare. The power of governments is reduced to a minimum in a world where Walmart accumulates in 1 year an amount of money ($166 billion) that is three times the whole GDP of Bangladesh. Whilst in 1970 85% of international commercial transactions were related to the real economy of goods and services, 20 years later, the percentage of transactions of a non-speculative type had been reduced to 2% of the total. The spectre of default keeps governments tuned into the demands of speculative capital. Multinational corporations also tend to pay few taxes: according to "Fortune", of the 288 most profitable companies in

2008–2012, 26 did not pay U.S. federal taxes (DeVogli 2012; http://tinyurl. com/kwpfkaz). How can we expect people to be motivated to pay taxes to contribute to welfare in this scenario where the irresponsible private appropriation of great wealth dominates? If Europe were to go into default and the political philosophy—based on the central role of welfare—that has supported it were to be dissolved, one of the consequences could be the privatisation of a part of the national health services and the introduction of a two-speed system: private insurance for the rich and an impoverished and low-quality public service for the poor, underfinanced by taxes.

A further component that is worth mentioning because of its extension is the endemic corruption throughout the entire financial system and in national and subnational governments, particularly in developing countries and in the health sector.

The Precautionary Principle

As the planet is facing terrible challenges linked to globalisation (the main one being climate change), it is fundamental to recognise all the early signs of crisis or deterioration, in particular from the point of view of public health. Observing and remembering the past can help to understand the consequences of underestimating the early warning signs. In the early 1960s, Europe was struck by a severe epidemic of malformations (phocomelia) caused by the use of the drug Thalidomide in pregnancy. In the United States, there was however a much more limited number of cases because the pharmacologist Frances Kelsey delayed the approval of the drug by the Food and Drug Administration based on its suspected teratogenicity.

Kelsey and the German doctors who described the epidemic of phocomelia in Germany underwent strong pressure from the pharmaceutical company Richardson-Merrell and were harshly criticised by colleagues because their observations were considered "preliminary". However, the epidemic was real and had enormous dimensions (between 10,000 and 20,000 cases): what would have happened if the fears had been completely hushed up with the complacency of some scientists? This affair shows the importance of early warning signs and the importance of the moral integrity of researchers, including in the face of personal risk. In reward for her integrity, Ms Kelsey was awarded an important civil decoration by John Kennedy.

The precautionary principle, which is included in the European Directives and in national laws, is based on two general criteria: (1) adequate action should be undertaken in response to evidence which is still limited, but plausible and credible, of damage to health and the environment, and (2) the burden of proof is shifted from demonstrating the presence of a risk to the demonstration of its absence (see the European Environmental Agency monograph, Late Lessons from Early Warnings, http://tinyurl.com/bxwfwls).

The good reasons to adopt the precautionary principle include, for example, the difficulty of establishing thresholds for the action of carcinogens; the long periods of latency for many non-communicable degenerative diseases; the uncertainties on the molecular mechanisms of action; and the widespread diffusion of new environmental contaminants, such as nanoparticles. There are also practical limits to implementing the precautionary principle: in particular, what is the minimum level of suspicion that leads to the application of the principle? The precautionary principle, if taken literally, may be paralysing; the fact that risks are evaluated separately from the benefits, in the case of drugs, vaccines and other products of public utility, can lead to decisions that are, at the very least, unilateral.

One of the reasons to take early warning signs seriously is the limited understanding of many diseases, probably influenced by environmental exposures but for which scientific evidence is still uncertain (Alzheimer, Parkinson, amyotrophic lateral sclerosis and many types of cancers). Associating the precautionary principle—understood reasonably—with more extensive and advanced scientific research on the environmental causes of diseases should allow protecting the population from new and old risks.

The interface between science and politics—of which the precautionary principle is a pivotal element—is one of the most difficult challenges of the future, which presents the looming threat of a limited availability of food and water in large areas of the world, mass migration and climate change. The role of the international agencies should be enormously reinforced in the coming years, at the risk of literally collapsing in the face of these challenges.

Environmental Sustainability

The sustainability of a biological species in a certain environment is measured by the maximum size of a population that that environment can sustain for an indefinite period of time, considering the reserves of food, water and other necessary assets available in it. It is also defined as the "maximum load" that can be supported by the environment. The productive activities of humans have a repercussion today on the planet which has never been experienced before, considering the dimensions of the contaminating activities or those which impoverish the resources. One of the best known examples is the effect of the acid rains on ecosystems. The phenomenon originated due to the type of predominant atmospheric pollution at the time (from the 1950s onwards), rich in sulphur dioxide, which reacting with the water in the atmosphere produced sulphuric acid. Having exceeded the capacity of the ecosystem to absorb and eliminate the sulphuric acid, the acid rains had important effects, especially on forests and fish in the lakes through the acidification of the waters. The phenomenon was considerably reduced in developed nations, with the reduction of sulphur dioxide in emissions, but it is now spreading in Russia and China.

It has been estimated that to support the impact of the human population, about 10 hectares per person in the rich countries, but only 2.5 hectares at a global level are necessary. Based on this and other measurements, there is therefore not enough land to support the world population at the level of the consumption of industrialised nations. However, obviously, nobody is enthusiastic at the idea of reducing consumption. On the contrary, there are widespread aspirations to increase industrial production and consumption in LMIC to be able (perhaps) to reduce poverty. Various attempts have been made to estimate how many people can live comfortably and sustainably on the planet, on the basis of some reasonable compromises between different standards of living. Realistic estimates based on the production of food, use of water, consumption of energy and a reduced impact of the carbon footprint stand at around 3–5 billion people, with a large range of different estimates. With a population of 7 billion and projections of up to 10 billion, the capacity of the planet to sustain the impact has probably already been amply exceeded. The trouble with these estimates is that *a posteriori* they prove sometimes completely wrong. For example, the size of the human population has doubled from the early 1970s, but there was no "Malthusian" catastrophe due to large increases in agriculture productivity and food availability (Deaton 2015).

On the Conflict of Interests

Effective prevention of climate change, of environmental deterioration and of their effects on health necessarily entails a fight against the conflicts of interests, as the "deniers" practically all work for large industry (the oil industry in the case of climate change). It is important however to clarify what is meant by conflict of interests, as there is a wide range of definitions, from the most simple and direct ones such as being employees of a public institution and at the same time being paid by the oil industry to study climate change to the more complex and indirect ones such as holding shares—maybe without even knowing it—in an industry that contributes to an increase in the levels of CO_2. On this theme, there are extensive cultural differences between the different countries: in the United States, the activity of lobbying is common and even the phenomenon of the revolving doors—businessmen who enter politics and then return to private business—is fully accepted. In Europe, on the other hand, some maintain that the fact that the International Agency for Research on Cancer (IARC) admits in its Monographs observers of industry (without the right of vote) is already a conflict of interest which invalidates the evaluative process (an opinion I do not agree with).

There are also great differences at European level. My experience in two government bodies, one in Italy and one in England, has been instructive. Nobody on the Consiglio Superiore di Sanità (Higher Health Council) was asked (at least, as long as I was a member of it) to declare any conflicts of interest, even though many of the resolutions—for example the approval of new postgraduate schools—were the result of negotiations between corporations (orthopaedics, plastic surgeons, nurses, etc.) and the outcome was conditioned by the position of each player at the end of the process. On the Committee on the Carcinogenicity of Chemicals of the British Department of Health, there is apparently even more conflict of interest because there are also representatives from industry, but everyone has to declare any conflicts of interest depending on the substance of which the carcinogenicity is being evaluated. Above all, the process is guided by scientific procedures; therefore, the interest of each party with respect to the outcomes is secondary with respect to a balanced evaluation of the evidence. To put it briefly, in England—at least a few years ago—the evidence took precedence over interests, in Italy, interests over the evidence (using the term interest in a very broad sense, from economic to corporative).

In my opinion, neither of the two models is ideal. With respect to both, the model of the IARC is certainly preferable: rigorous procedures, evidence-based decisions and transparency of conflicts of interest with a role exclusively as observers for the industrial representatives. The model of the IARC reflects what in my opinion is one of the highest principles established to guarantee the functioning of a democracy, the "veil of ignorance" proposed by John Rawls: those who are summoned to decide on a public question must not be influenced by the position that they will occupy as a consequence of the decision, i.e. they must decide behind a veil of ignorance. In other words, the deciders are imagined to have to make their decisions ignoring both the position they occupy at the time and the position they will occupy afterwards: only when the veil is lifted will they know in which way that decision modifies their position and their assets.

These subjects have been widely taken up—with significant differences—by the greatest representative today of public ethics, Michael Sandel. We will talk about him again in the last chapter.

An implicit definition of conflict of interests emerges from the previous considerations, which can be made clear as follows: "*A conflict of interest is a set of circumstances that creates a risk that professional judgment or actions regarding a primary interest will be unduly influenced by a secondary interest*" (definition by Lo and Field 2009). In the case of public health, the primary interest is the health of the population, whilst secondary interests can be profit or career, as shown by the case of Thalidomide. The moral integrity of the scientist consists of recognising and avoiding the conflict of interests.

The conflict of interests is not exclusively a result of corruption but is much more widespread than we believe. It is often blurred and its regulation ought to be entrusted to practice and not only to regulations. For example, in a research facility, the fact that a Director of a Department uses material and human resources mainly for the purposes of his own laboratory, rather than being impartial, represents a conflict of interests. The emphasis by the European Union on the marketability of products of research and on the Community impact, understood as an economic impact, is pertinent to the conflict of interests. In the case of climate change, for example, it is possible that many of the technological solutions aimed at reducing it or preventing it have an initially negative economic impact and are not translated into immediate economic advantages: the insistence on the economic repercussions risks distorting the choice and plan of investigations which—left to the sole rationality of the research—could go in an opposite direction. And to mention other examples, there are spin-offs of universities created to market the products of research where at times what has been discovered does not have the virtues that are advertised. One example is the kit to identify people who find it hardest to stop smoking, based on the DRD2 gene, for which there are contrasting evaluations between scientific articles and the commercial promises (Munafò et al. 2009).

The world of communication is also affected by conflict of interests. The insistence on "parity of terms" (*par condicio*), which developed in a period of great political and media disagreement, is one example: it has often been used to give dignity to positions such as climate denialism which should deserve far less publicity. Parity of terms contributes to creating smokescreens—activities in which institutions created by large industry such as the Cato Institute (http://www.cato.org/) are literally specialised—and to fuelling the needs for scandal and drama of the information media.

Until these problems are clarified at global level, it is difficult to find convincing answers to challenges such as climate change. A purely regulatory-procedural approach to the conflict of interests is necessary but not sufficient and may not be effective. I find Sandel's approach more effective: it is based on the diffusion and internalisation of the concept of public ethics, through the creation of consensus in society. The problem is that there are now very few places where public ethics can be exercised and where its presuppositions and applications can be discussed. This trend is very well described by Zygmunt Bauman: we face the transition to liquid societies where there are no longer the three levels of microstructure (the family), mesostructure (political parties, trade unions and movements) and macro-structure (society as a whole, the economy). The "great transformation" of the environment and of public health is taking place in parallel with the substantial disappearance of the mesostructures, replaced by the direct relationship between individuals through the Internet and social media and by the hyper-trophy of megastructures with the absolute dominance of the economy and finance.

All the consequences of this transformation have yet to be identified, but they have clearly marked the end of the twentieth century, that period when public ethics was discussed in the mesostructure (parties, Unions and Churches) and reflected on the microstructures (families). The greatest eco-logical crisis of humanity is taking place jointly with the greatest change in communication and decision-making ever. How the conflicts of interests are shaped through the Internet networks of a liquid society is everything but obvious.

The same concepts have been expressed simply but masterfully by the historian Tony Judt (2011, p. 121):

> Young people are indeed in touch with likeminded persons many thousands of miles away. But even if the students of Berkeley, Berlin and Bangalore share a common set of interests, these do not translate into *community*. Space matters. And politics is a function of space.

With regard to the erosion of respect for the public sphere (p. 129):

> And once we cease to value the public over the private, surely we shall come in time to have difficulty seeing just why we should value law (*the* public good par excellence) over force.

10

Public Health as a Common Good

In recent years, public health has had to face major challenges and moral dilemmas. As we have seen, the challenges have included economic crises, globalisation, climate change and their effects on health. The moral dilemmas concern a growing trend to understand health, including prevention, as a consumer good to be enjoyed individually.

The transgenerational transmission of culture goes hand in hand with the transmission of the "healthy" genes between generations. The cultural continuity between generations—between past, present and future—is necessary for the transfer of knowledge, for moral values and for social cohesion. It therefore implies for the past and future generations the same respect that we have for the present one. Can this be said to be happening? Utilitarian thought has often put the accent above all on the people living at the present time and the ethics of the obligations towards future generations has not yet fully emerged (nor has the importance of the historical memory).

This chapter opposes the classical liberal conception based on "individual freedom of choice" to a different idea of freedom based on individual and collective responsibility.

The dominant concept in the moral philosophy of the past three centuries has been the pursuit of happiness (Locke); however, this principle can be interpreted in many different ways. The predominant interpretation, which found complete expression in mature capitalism, is the conception of John Locke, according to whom the pursuit of happiness can be obtained in particular through private initiative, i.e. the private use of land and natural resources. In his *Two treatises of government* (1690), Locke describes the

© Springer International Publishing AG 2017
P. Vineis, *Health Without Borders*, DOI 10.1007/978-3-319-52446-7_10

laborious exploitation of natural resources as the only opportunity for humanity to draw benefit from the world God gave them, and to pursue happiness through the transformation of the environment. The original idea according to which the most effective pursuit of happiness is allowed by individual initiative and by public appropriation is at the root of some of the moral dilemmas facing us today, which come from what we could call the "individualist turn".

Locke's inheritance was consolidated and exasperated by the spread of neo-liberalism in the 1970s, which further scaled down the notion of public assets. On the cultural level, neo-liberalism is now spreading to the low-income countries, with the emerging middle-classes dominated by compulsive consumerism. This tendency is translated at the level of government/institutions in the individualised promotion of health rather than in structural actions to prevent disease; in the at times uncritical spread of predictive medicine (including genetic tests); in the privatisation of water; and even in the patentability of genes and parts of the body, a phenomenon which is not comprehensible outside this context.

Madison Against Jefferson

The patenting of genes is a very controversial topic. After identifying the gene for the predisposition to breast cancer—BRCA1—the private company Myriad patented the DNA sequence, thus obtaining the commercial rights on all the clinical applications deriving from the use of the test. In 2009, a group of doctors and women affected by breast cancer contested the fact that the DNA of BRCA1—and by implication all DNA—could be patented.

The practice of patenting has a long history, in particular in the United States, where it is based on ideas and debates that go back to the American Revolution. Thomas Jefferson opposed monopolies deriving from patents and from the "copyright" system for a long time, but James Madison persuaded him of their value as an incentive for authors and inventors, to the extent that a patent was temporary (Kevles 2013). Jefferson, accepting the general idea, nevertheless emphasised the requisite of the innovation and of the added value, i.e. the need for an innovative component; in practice, the constituents of nature such as the elements in the periodic table could not be patented.

The discussion on the Myriad case developed around the notions of added value and invention, between those who maintained that isolating a sequence of DNA is simply the discovery of a law of nature (and therefore belongs to the whole of humanity) and those who consider it an invention and therefore the property of the inventor. When the Myriad case was brought before the U.S. Supreme Court, the judge Stephen Breyer decided that the relationship between the sequence of DNA and the disease could not be patented because it was a law of nature. Myriad opposed to this interpretation the fact that the "discovery" lays in the development of the clinical methods and tests. After various phases of appeal, the final act was the decision of the Supreme Court in June 2013, which definitively excluded that the gene could be patented as such. Making up for this, it left a technical loophole allowing the patent of the cDNA, or the complementary DNA, i.e. a copy of the gene after its transcription into mRNA. These are the "*problems of nature in the time of its technical reproducibility*", to paraphrase Walter Benjamin.

The story of Myriad shows that there is a long tradition of thought that tends to protect common assets, i.e. the right of the whole of humanity to have access to the products of scientific discoveries (DNA and the relative common assets such as the diagnostic tests) and to preserve the integrity of nature (a common asset in itself) independently of the commercial exploitation.

Mill and the "Harm Principle"

One subject that is apparently completely different is primary prevention through the interference of the State with individual choices, which is contemptuously defined "the role of the nanny-State". When the former Mayor of New York, Bloomberg, proposed banning the sale of sweetened drinks in maxi-sizes in 2012, in an attempt to block the obesity epidemic, many in the United States—for example the Bureau of Consumer Protection—contested this decision, appealing to the central principle of the liberal States expressed by John Stuart Mill in his *On Liberty*. For Mill, as a general rule, a government cannot exercise coercion on the individual to protect her from herself. The only purpose for which power can be exercised on any member whosoever of a civilised community, against her will, is to prevent harm to others (known as the "harm principle"). In other words, at individual level, each person has the right to judge what is good or bad for them.

Alongside the argued and meditated positions, some made real caricatures of the work of Bloomberg, like the famous English journalist, Christopher Hitchens who, in an article in "Vanity Fair" in 2004, wrote: ". . . one of the world's most broad-minded and open cities is now in the hands of a picknose control freak." (See Magnusson 2014 and articles linked to the Bloomberg case.) It appears that Hitchens' hostility was in part linked to a specific episode, when the editor of "Vanity Fair" reported having undergone a search and a fine for having smoked in the magazine's office and then only for the fact that he had an ashtray on his desk.

Amongst the counter-arguments to the objections of critics, one important theme is that individual choice is limited by the availability of goods, their costs and advertising and marketing policies. For example, the promotional campaign of Frito-Lay in Thailand was followed by a strong consumption of junk food by children and youngsters (Hawkes et al. 2009). There are also many types of behaviour linked to health that have consequences on others or on public assets. The use of heroin, for example, fuels crime and violence, and alcohol abuse has numerous negative consequences, not least road accidents. The increase in diseases linked to obesity is beginning to overcrowd many public hospitals, already at the limit of their capacities. Fundamentally, it is extremely difficult to draw an exact line between purely individual behaviour which does not have consequences on others and those which, possibly indirectly, do.

Reducing the debate to the dichotomy between individual choice and the nanny-State is a hyper-simplification, which does not consider the disproportion of power between the individual and an oligopolistic food industry which did not exist in Mill's time.

In a recent and provocative book (Sunstein 2013; Conly 2013), Conley argues that, for State paternalism to be justified, a number of criteria have to be satisfied, including a reasonable expectation that the action is effective and the benefits exceed the costs. Applying these criteria, Conley deems that banning trans fatty acids in New York is an excellent example of justified coercion. On the basis of scientific evidence, the ban has been effective in bringing about immediate benefits for public health, and those benefits are much greater than the costs. Conley also agrees with the measures aimed at reducing food portions, again in order to prevent the obesity epidemic.

Freedom of Choice, Freedom of Enterprise

What do State paternalism and the patenting of genes have in common? In both cases freedom is at stake: that of consumers and that of industry. On the front of the consumers, the classic liberal argument is that the individual is free to choose on condition that his choice does not harm others. On the side of industry, the liberal argument is that the investment and the risks involved in research and development justify the patents and the protection of intellectual property. But in the case of genes, it was not an "invention" clearly made by humans, rich in terms of intellectual property and innovation, but the "discovery" of parts of nature, like DNA sequences.

The patent was defended by calling into question freedom of choice: the argument used is that the consumer would not be able to exercise his own free choice without the intervention of industry, which for example makes the genetic test available; in other words, the commodification of nature makes nature available to the consumer. However, the usefulness of a test, in the context of the free market, can be widely overestimated and advertised by the manufacturer only to increase sales. This is not the case of BRCA1, the usefulness of which is undisputed, but the commercialisation of genetic tests to identify people more susceptible to the action of carcinogens or less inclined to stop smoking, already a reality today, is far more questionable (Vineis and Christiani 2004; Vineis et al. 2005).

An example of how misleading low-penetrance genetic tests can be is the proposal to select smokers who have the greatest difficulty in stopping, to treat them in a more pharmacologically aggressive way. The tests available, based for example on the DRD2 gene, poorly discriminate dependence on smoking, with too great an error of classification for the tests to be really useful. Other proposals of application of genetic tests to the case of smoking have concerned the susceptibility of smokers to the disease instead of the neurological phenomenon of dependence. There are genes (like GSTM1, CYP1A1 and many others) the variants of which increase the susceptibility to cancer in smokers without acting through dependence.

As a possible effect, and which in part has been proven, there is the risk that those with the "normal" or commoner variant feel protected and do not stop smoking. In other words, the effect of smoking is far greater than that of the gene variants and stopping smoking is far more effective than any apparent protection exercised by gene variation (Sanderson et al. 2009). In conclusion, genetic tests based on low-penetrance gene variants do not have solid scientific bases and represent a distorted use of the resources, even though they are part of the "innovative" proposals of what is known as stratified or personalised medicine/prevention, in the name of free choice.

There is a strong tendency to promote individualised ways to prevent non-communicable diseases, even though they have been seen to be ineffective in randomised experimentation (Ebrahim and Smith 2001). The highly publicised 25 × 25 strategy (reduce mortality by chronic diseases by 25% by 2025) unfortunately places a great deal of emphasis on individual risk factors rather than on the objectives at the level of population and interventions of the State. The institutional initiatives, like banning trans fatty acids or taxation of harmful substances, remain by far the most effective path, as clearly shown by the cases of alcohol and cigarettes even though they violate Mill's harm principle, according to which it cannot be the State that establishes what is good or bad for the individual.

An interesting example of the contrast between private vs. public initiatives bearing on public health is the different responsibilities for sewage waste piping in California and in most European countries. In California, the individual is responsible for all the piping from their house to the main sewer pipe. In Europe, the municipality is responsible for all the pipework, with the home owner only responsible for connections from household appliances to the lateral pipework. This has huge potential impacts on water-borne diseases for a community (Shah Ebrahim, personal communication).

More generally, it is critical to separate health protection from health promotion and other preventive activities. Health protection requires collective activities, in general through legislation, taxation or common values. Health promotion at the individual level—as currently conducted in the health systems—is less effective, and it is unlikely to have major impacts on disease or mortality rates.

Climate Change and Public Ethics

Climate change is a paradigmatic case of the moral dilemmas arising in public health for various reasons. In the first place, the effects on health may be multiple, even dramatic, but they are still poorly known. Alongside obvious phenomena like heat waves or floods, there are also less obvious effects such as the spread of infectious diseases, the diminishing availability of food and water of good quality and the increase of migrations due to sea level rise or drought. A second reason is the fact that preventing climate change and its consequences can be done only marginally at individual level. Most of the measures have to be taken collectively and precisely in the parts of the world that will be less subject to undergoing the effects of climate change. The third point is that there are considerable doubts on the actions that can prevent climate change in the most effective way.

One of the best known measures was the so-called cap and trade, i.e. the opportunity offered to companies in the richest countries, and that produce a greater amount of greenhouse gas, to buy the permit to issue CO_2 from low-income countries, in addition to (or in place of) reducing their own emissions. One motivation can consist in the fact that reduction can require radical investments for which not all companies are prepared and as a temporary measure they can therefore buy the rights to emissions from emerging companies in low-income countries.

Although from an economic point of view the proposal is rational and effective, the ethicist Michael Sandel (2005) contests it on the grounds of a moral argument: the cap and trade suggests that it is possible to negotiate pollution and climate change (and nature by extension), as sub-products of industrial society, i.e. that they can be subjected to the laws of the market. In addition, according to Sandel, the cap-and-trade policy damages international cooperation and the sense of responsibility, which are the only ways to effectively tackle climate change in the long term.

The distinction between assets (common goods) and rights is central in Sandel's philosophy, because it expresses the contrast between the emphasis on individual rights of appropriation (of health and education, for example) and the identification of the purposes and defence of common assets (Sandel 2012). In the words of Tony Judt: "To convince *others* that something is right or wrong, we need a language of ends, not means" (Judt 2011, p. 180). The language of ends is the language of politics which today is wholly dominated by the utilitarian discourse of the means, i.e. of the economy. The language of the ends is that of the ethics of freedom understood in the Kantian sense as adhesion to a high ideal of justice, whilst the language of the means is the language of freedom understood individualistically as pure freedom to express one's Ego fully and selfishly (Sandel 2010).

Conclusions

Today, it seems that there is a trend in medicine to promote freedom of choice and individualised access to the "health asset", including genetic tests, the tests of predictive medicine and the patenting of natural discoveries. There are numerous negative repercussions of this trend which the economic crisis has dramatically brought into light.

In the first place, the solely individual promotion of health has not proven to be effective, except in very limited contexts, and leads to increased social inequality (Marmot et al. 2012). It is surprising that some eminent epidemiologists a few years ago maintained, in a famous article in "The Lancet" (Rothman et al. 1998), that the task of epidemiologists is not to deal with poverty but only to discover the causes of diseases. Even for a subject that seemed remote at the time such as climate change, poverty is not only a decisive modifier of the effects of other risk factors, but is itself one of the most important "causes of the causes".

The great successes in the fight against infections in the nineteenth century derived from collective actions, in particular the introduction of the sewage system, drinking water, town planning and vaccinations. Major investments were also made in the twentieth century and enormous successes were achieved in the fight against transmissible diseases, in particular against smallpox, the mass vaccinations against hepatitis B and measles, and today the GAVI alliance. However, we are not seeing similar systematic efforts for non-communicable diseases, except through the taxation of cigarettes and alcohol, the prohibition of trans fatty acids in New York and some other sporadic examples.

In the second place, the growing private appropriation of nature (land, water and, fortunately, for the time being, not genes. . .) incorporates a process of commodification which, far from being virtuous in the sense of the interpretation by Madison and Locke, opens up the path to speculative financial operations as has already taken place for other assets. The first consequences have already been seen; in recent years, dietary deficiencies and even famines in various parts of the world have depended on fluctuations of value of the crops on the stock market. Once the hedge funds or futures are also applied to food or to genes, the situation could go completely out of control.

Lastly, there are challenges such as climate change which, by definition, cannot be tackled without a joint international effort. One of the most dangerous attitudes we are seeing is denial: denial of the real roots of the financial crisis and denial of its consequences for health; denial of climate change and denial of the extent of the (partly irreversible) contamination of the planet. The problem of denial intersects with that of the conflict of interests, i.e. the trend, apparently on the increase, of scientists giving up their independence and providing arguments that are useful to the bearers of economic interests.

An important point of view on the common good is that of Michael Sandel, who describes many examples of the market's intrusion into sectors that should preserve their specificity. For example, according to Sandel American schools are increasingly at the centre of a mechanism of private donations and sponsoring with advertising aims. However, as Sandel writes, school exists not to encourage consumption, but rather to educate to consider one's aims in life—consumption included—in a critical and significant pattern. Education is an end in itself, and not only a means. In the United States and increasingly in the rest of the world, education is perceived exclusively as a vehicle towards the labour market and this perception is accompanied by a claim for social promotion. School is naturally first of all an institution that "promotes non-commercial ideals", such as the pursuit of truth and moral and civil sensitivity. Sandel's criticism also extends to biomedical technologies. We have to reflect whether by entrusting the potential of biomedical technologies only to market forces we want to aim more for the individual skill of adaptation and self-promotion or reinforcing everyone, including the less fortunate.

To Know More

I have deliberately left out many specialised references in the bibliography. However, I suggest further reading or consulting materials such as the invaluable videos of Gapminder (www.gapminder.org).

Contents Available Online

CO_2 emissions in the world: http://tinyurl.com/o5blhhj
Epidemic of lung cancer: http://tinyurl.com/lk5w9ut
Infantile mortality in Bangladesh and in the rest of the world: http://tinyurl.com/qzabkxt
Obesity and the built environment: The Weight of the World, http://tinyurl.com/kxz72fk

For a **Glossary** of technical terms used in the text (particularly in epidemiology and public health), I suggest: Miquel Porta, A Dictionary of Epidemiology. Oxford University Press, Oxford, 2014 (http://www.oxfordreference.com/view/10.1093/acref/9780195314496.001.0001/acref-9780195314496)

© Springer International Publishing AG 2017
P. Vineis, *Health Without Borders*, DOI 10.1007/978-3-319-52446-7

Health Without Borders: Epidemics in the Era of Globalisation. Climate Change (Jakob Levi)

(2017) Shot in Bangladesh and London, and assembled with a collage of found footage, digital and 16mm animation. This video is to accompany the launch of the new book by Professor Paolo Vineis called Health Without Borders: Epidemics in the Era of Globalisation. This film only focuses on the issue of climate change, summarising much of the science and drawing on many examples from Bangladesh.

Recommended Reading

IPCC Fifth Assessment Report (AR5): http://ipcc.ch/report/ar5/

McMichael, A. J. (2001). *Human frontiers, environments and disease: Past patterns, uncertain futures.* Cambridge, MA: Cambridge University Press.

Skolnick, R. (2008). *Essentials of global health.* Sudbury, ON: Jones and Bartlett Publ.

Butler, B., Jane Dixon, J., & Capon, A. (Eds.). (2015). *Health of people, places and planet.* Canberra: Australian National University Press.

Bibliography

Alleyne, G., Binagwaho, A., Haines, A., Jahan, S., Nugent, R., & Rojhani, A. (2013). Stuckler D; Lancet NCD Action Group. Embedding non-communicable diseases in the post-2015 development agenda. *Lancet, 381*(9866), 566–574. doi:10.1016/S0140-6736(12)61806-6.

Atkinson, A. B. (2015). *Inequality. What can be done?* Cambridge, MA: Harvard University Press.

Baker, P., Kay, A., & Walls, H. (2014). Trade and investment liberalisation and Asia's noncommunicable disease epidemic: A synthesis of data and existing literature. *Globalization and Health, 10*, 66. doi:10.1186/s12992-014-0066-8.

Beelen, R., et al. (2014a). Effects of long-term exposure to air pollution on natural-causrtality: An analysis of 22 European cohorts within the multicentre ESCAPE project. *Lancet, 383*(9919), 785–795. http://tinyurl.com/ltn4ymg

Beelen, R. et al. (2014b). Long-term exposure to air pollution and cardiovascular mortality: An analysis of 22 European cohorts. *Epidemiology, 25*(3), 368–378.

Bleich, S. N., et al. (2012). Health inequalities: Trends, progress, and policy. *Annual Review of Public Health, 33*, 7–40. http://tinyurl.com/mh52lb4

Bloom, D. E., & Canning, D. (2007). Commentary: The Preston Curve 30 years on: Still sparking fires. *International Journal of Epidemiology, 36*(3), 498–499. Discussion 502-3.

Boffetta, P., et al. (2011). TCDD and cancer: A critical review of epidemiologic studies. *Critical Reviews in Toxicology, 41*(7), 622–636. Study financed by the American Chemistry Council.

Bourguignon, F. (2012). *La mondialisation de l'inégalité.* Paris: Seuil.

Bray, F., Jemal, A., Torre, L. A., Forman, D., & Vineis, P. (2015). Long-term realism and cost-effectiveness: Primary prevention in combatting cancer and associated

© Springer International Publishing AG 2017
P. Vineis, *Health Without Borders*, DOI 10.1007/978-3-319-52446-7

inequalities worldwide. *JNCI: Journal of the National Cancer Institute, 107*(12), djv273.

Bray, F., et al. (2012). Global cancer transitions according to the Human Development Index (2008-2030): A population-based study. *The Lancet Oncology, 13,* 790–801.

Brouwer, R., et al. (2007). Socioeconomic vulnerability and adaptation to environmental risk: A case study of climate change and flooding in Bangladesh. *Risk Analysis, 27*(2), 313–326.

Case, A., & Deaton, A. (2015). Rising morbidity and mortality in midlife among white non-Hispanic Americans in the 21st century. *Proceedings of the National Academy of Sciences of the United States of America, 112*(49), 15078–15083.

Chang, H.-J. (2010). *23 things they don't tell you about capitalism.* London: Penguin Books.

Chen, Y., & Ahsan, H. (2004). Cancer burden from arsenic in drinking water in Bangladesh. *American Journal of Public Health, 94,* 741–744.

Clapp, J. (2001). *Toxic exports: The transfer of hazardous wastes from rich to poor countries.* Ithaca, NY: Cornell University Press.

Cogliano, V. J., et al. (2011). Preventable exposures associated with human cancers. *Journal of the National Cancer Institute, 103*(24), 1827–1839.

Cokkinides, V., et al. (2009). Tobacco control in the United States: Recent progress and opportunities. *CA: A Cancer Journal for Clinicians, 59*(6), 352–365.

Conly, S. (2013). *Against autonomy: Justifying coercive paternalism.* Cambridge, MA: Cambridge University Press.

Cotty, P. J., & Jaime-Garcia, R. (2007). Influences of climate on aflatoxin producing fungi and aflatoxin contamination. *International Journal of Food Microbiology, 119* (1–2), 109–115.

Davis, J., & Tallis, R. (2013). *NHS SOS.* London: Oneworld.

De Martel, C., et al. (2012). Global burden of cancers attributable to infections in 2008: A review and synthetic analysis. *The Lancet Oncology, 13*(6), 607–615.

De Rooij, S. R., et al. (2007). The metabolic syndrome in adults prenatally exposed to the Dutch famine. *The American Journal of Clinical Nutrition, 86*(4), 1219–1224.

De Vogli, R. (2012). *Progress or collapse—The crises of market greed.* London: Routledge/Taylor and Francis.

Deaton, A. (2015). *The great escape. Health, wealth, and the origins of inequality.* Princeton, NJ: Princeton University Press.

Demetriou, C. A., et al. (2013). Methylome analysis and epigenetic changes associated with menarcheal age. *PLoS One, 8*(11), e79.391.

Diderichsen, F., Evans, T., & Whitehead, M. (2001). The social basis of disparities in health. In M. Whitehead et al. (Eds.), *Challenging inequities in health: From ethics to action.* New York: Oxford University Press.

Dixon, J., Banwell, C., Seubsman, S. A., Kanponai, W., Friel, S., & Maclennan, R. (2007). Dietary diversity in Khon Kaen, Thailand, 1988 2006. *International Journal of Epidemiology, 36*(3), 518–521.

Doll, R., & Peto, R. (1981). The causes of cancer: Quantitative estimates of avoidable risks of cancer in the United States today. *Journal of the National Cancer Institute, 66*(6), 1191–1308.

Ebrahim, S., & Smith, G. D. (2001). Exporting failure? Coronary heart disease and stroke in developing countries. *International Journal of Epidemiology, 30*(2), 201–205.

Eheman, C., et al. (2012). Annual report to the nation on the status of cancer, 1975-2008, featuring cancers associated with excess weight and lack of sufficient physical activity. *Cancer, 118*(9), 2338–2366.

Farmer, P., et al. (2010). Expansion of cancer care and control in countries of low and middle income: A call to action. *Lancet, 376*(9747), 1186–1193.

Friel, S., Dangour, A. D., Garnett, T., Lock, K., Chalabi, Z., Roberts, I., Butler, A., Butler, C. D., Waage, J., McMichael, A. J., & Haines, A. (2009). Public health benefits of strategies to reduce greenhouse-gas emissions: Food and agriculture. *Lancet, 374*, 2016–2025.

Gallino, L. (2013). *Il colpo di stato di banche e governi.* Torino: Einaudi.

Gallo, V., et al. (2012). Social inequalities and mortality in Europe. Results from a large multi-national cohort. *PLoS One, 7*(7), e39013. http://tinyurl.com/orr5wkb

Gluckman, P., & Hanson, M. (2006). *Mismatch. The lifestyle diseases timebomb.* Oxford: Oxford University Press.

Gordon, S. B., Bruce, N. G., Grigg, J., Hibberd, P. L., Kurmi, O. P., Lam, K. B., Mortimer, K., Asante, K. P., Balakrishnan, K., Balmes, J., Bar-Zeev, N., Bates, M. N., Breysse, P. N., Buist, S., Chen, Z., Havens, D., Jack, D., Jindal, S., Kan, H., Mehta, S., Moschovis, P., Naeher, L., Patel, A., Perez-Padilla, R., Pope, D., Rylance, J., Semple, S., & Martin II, W. J. (2014). Respiratory risks from household air pollution in low and middle income countries. *The Lancet Respiratory Medicine, 2*(10), 823–860. doi:10.1016/S2213-2600(14)70168-7.

Haines, A., McMichael, A. J., Smith, K. R., Roberts, I., Woodcock, J., Markandya, A., Armstrong, B. G., Campbell-Lendrum, D., Dangour, A. D., Davies, M., Bruce, N., Tonne, C., Barrett, M., & Wilkinson, P. (2009). Public health benefits of strategies to reduce greenhouse-gas emissions: Overview and implications for policy makers. *Lancet, 374*(9707), 2104–2114. doi:10.1016/S0140-6736(09)61759-1.

Hawkes, C., et al. (2009). Globalisation, trade and the nutrition transition. In R. Labonté et al. (Eds.), *Globalisation and health: Pathways, evidence and policy.* New York: Routledge.

Heard, E., & Martienssen, R. A. (2014). Transgenerational epigenetic inheritance: Myths and mechanisms. *Cell, 157*(1), 95–109. doi:10.1016/j.cell.2014.02.045.

Heijmans, B. T., et al. (2008). Persistent epigenetic differences associated with prenatal exposure to famine in humans. *Proceedings of the National Academy of Sciences of the United States of America, 105,* 17.046–17.049.

Hessel, P., Vandoros, S., & Avendano, M. (2014). The differential impact of the financial crisis on health in Ireland and Greece: A quasi-experimental approach. *Public Health, 128*(10), 911–919. doi:10.1016/j.puhe.2014.08.004.

Jenkins, R. (2004). Globalisation, production, employment and poverty: Debated and evidence. *Journal of International Development, 16,* 1–12.

Judt, T. (2011). *Ill fares the land.* London: Penguin Books.

Karamanoli, E. (2015). 5 years of austerity takes its toll on Greek health care. *Lancet, 386*(10010), 2239–2240.

Karanikolos, M., et al. (2013). Financial crisis, austerity, and health in Europe. *Lancet, 381*(9874), 1323–1331. http://tinyurl.com/lm8hh6n

Kelley, C. P., Mohtadi, S., Cane, M. A., Seager, R., & Kushnir, Y. (2015). Climate change in the Fertile Crescent and implications of the recent Syrian drought. *Proceedings of the National Academy of Sciences of the United States of America, 112*(11), 3241–3246.

Kentikelenis, A., Karanikolos, M., Papanicolas, I., Basu, S., McKee, M., & Stuckler, D. (2011). Health effects of financial crisis: Omens of a Greek tragedy. *Lancet, 378* (9801), 1457–1458.

Kentikelenis, A., et al. (2014). Greece's health crisis: From austerity to denialism. *Lancet, 383*(9918), 748–753. http://tinyurl.com/nlphjnc

Kevles, D. J. (2013, March 7). Can they patent your genes? *New York Review of Books.*

Khan, A. E., et al. (2011, April 12). Drinking water salinity and maternal health in coastal Bangladesh: Implications of climate change. *Environmental Health Perspectives.* http://tinyurl.com/otcoehq

Kirby, T. (2010). Canada accused of hypocrisy over asbestos exports. *Lancet, 376* (9757), 1973–1974.

Kontis, V., Bennett, J. E., Mathers, C. D., Li, G., Foreman, K., & Ezzati, M. (2017). Future life expectancy in 35 industrialised countries: Projections with a Bayesian model ensemble. *Lancet, 389*(10076), 1323–1335.

Krishnan, M., & Ray, S. G. (2010, August 7). Banned in 52 countries, asbestos is India's next big killer. *Tehelka Magazine,* 731. http://www.tehelka.com/story_main46.asp?filename=Cr070810banned.asp

Kucharski, R., Maleszka, J., Foret, S., & Maleszka, R. (2008). Nutritional control of reproductive status in honeybees via DNA methylation. *Science, 319*(5871), 1827–1833.

Labonté, R., et al. (Eds.). (2009). *Globalisation and health: Pathways, evidence and policy.* New York: Routledge.

Lansley, S., & Mack, J. (2015). *Breadline Britain. The rise of mass poverty.* London: Oneworld.

Lara, R. J., et al. (2009). Influence of catastrophic climatic events and human waste on Vibrio distribution in the Karnaphuli estuary, Bangladesh. *Ecohealth, 6*(2), 279–286. http://tinyurl.com/krfq7g3

Lewis, L., et al. (2005). Aflatoxin contamination of commercial maize products during an outbreak of acute aflatoxicosis in eastern and central Kenya. *Environmental Health Perspectives, 113*(12), 1763–1767. http://tinyurl.com/p28mnnb.

Lindert, J., et al. (2009). Depression and anxiety in labor migrants and refugees: A systematic review and meta-analysis. *Social Science and Medicine, 69*(2), 246–257. http://tinyurl.com/kyzbbc9

Lo, B., Field, M. J. (Eds.)., Institute of Medicine (US)., & Committee on Conflict of Interest in Medical Research, Education, and Practice. (2009). *Conflict of interest in medical research, education, and practice.* Washington, DC: National Academies Press.

Lovasi, G. S. (2009). Built environments and obesity in disadvantaged populations. *Epidemiologic Reviews, 31*, 7–20.

Lyko, F., et al. (2010). The honey bee epigenomes: Differential methylation of brain DNA in queens and workers. *PLoS Biology, 8*(11), e1.000.506.

Magnusson, R. (2014). Bloomberg, Hitchens, and the libertarian critique. *Hastings Center Report, 44*(1), 3–4.

Marmot, M., et al. (2012). WHO European review of social determinants of health and the health divide. *Lancet, 380*(9846), 1011–1029.

Martinez, L., et al. (2011). Impact of early developmental arsenic exposure on promotor CpG-island methylation of genes involved in neuronal plasticity. *Neurochemistry International, 58*, 574–581.

McClintock, B. (1951). Chromosome organization and genic expression. *Cold Spring Harbor Symposia on Quantitative Biology, 16*, 13–47.

McKay Illari, P., Russo, F., & Williamson, J. (Eds.). (2011). *Causality in the sciences.* Oxford: Oxford University Press.

Moodie, R., et al. (2013). Profits and pandemics: Prevention of harmful effects of tobacco, alcohol, and ultra-processed food and drink industries. *Lancet, 381*, 670–679.

Munafò, M. R., et al. (2009). Lack of association of DRD2 rs1800497 (Taq1A) polymorphism with smoking cessation in a nicotine replacement therapy randomised trial. *Nicotine and Tobacco Research, 11*(4), 404–407.

Nordhaus, W. (2012, March 22). Why the global warming skeptics are wrong. *New York Review of Books.*

Olshansky, S. J., et al. (2005). A potential decline in life expectancy in the United States in the 21st century. *The New England Journal of Medicine, 352*(11), 1138–1145.

Oreskes, N., & Conway, E. M. (2010). *Merchants of doubt.* New York: Bloomsbury.

Pallansch, M. A., & Sandhu, H. S. (2006). The eradication of polio – Progress and challenges. *The New England Journal of Medicine, 355*, 2508–2511.

Park, J., Hisanaga, N., & Kim, Y. (2009). Transfer of occupational health problems from a developed to a developing country: Lessons from the Japan-South Korea experience. *American Journal of Industrial Medicine, 52*(8), 625–632.

Parkin, D. M., Boyd, L., & Walker, L. C. (2011). The fraction of cancer attributable to lifestyle and environmental factors in the UK in 2010. *British Journal of Cancer, 105*(Suppl. 2), S77–S81.

Paterson, R. R. M., & Lima, N. (2010). How will climate change affect mycotoxins in food? *Food Research International, 43*(7), 1902–1914.

Pearce, N., Ebrahim, S., McKee, M., Lamptey, P., Barreto, M. L., Matheson, D., Walls, H., Foliaki, S., Miranda, J., Chimeddamba, O., Marcos, L. G., Haines, A., & Vineis, P. (2014). The road to 25×25: How can the five-target strategy reach its goal? *The Lancet Global Health, 2*(3), e126–e128. doi:10.1016/S2214-109X(14)70015-4.

PLoS Medicine. (Eds.). (2012). PLoS Medicine series on Big Food: The food industry is ripe for scrutiny. *PLoS Medicine, 9*(6), 1–2.

Probst, C., Njapau, H., & Cotty, P. J. (2007). Outbreak of an acute aflatoxicosis in Kenya in 2004: Identification of the causal agent. *Applied and Environmental Microbiology, 73*(8), 2762–2764.

Raaschou-Nielsen, O., et al. (2013). Air pollution and lung cancer incidence in 17 European cohorts: Prospective analyses from the European Study of Cohorts for Air Pollution Effects (ESCAPE). *The Lancet Oncology, 14*(9), 813–822.

Richmond, R. C., Davey Smith, G., Ness, A. R., den Hoed, M., McMahon, G., & Timpson, N. J. (2014). Assessing causality in the association between child adiposity and physical activity levels: A Mendelian randomization analysis. *PLoS Medicine, 11*(3), e1001618.

Rodrik, D. (2011). *The globalisation paradox*. Oxford: Oxford University Press.

Rogers, D. J., Randolph, S. E., Snow, R. W., & Hay, S. I. (2002). Satellite imagery in the study and forecast of malaria. *Nature, 415*(6872), 710–715.

Romaguera, D., et al. (2012). Is concordance with World Cancer Research Fund/American Institute for Cancer Research guidelines for cancer prevention related to subsequent risk of cancer? Results from the EPIC study. *The American Journal of Clinical Nutrition, 96*(1), 150–163.

Ross, R. K., Yuan, J. M., Yu, M. C., Wogan, G. N., Qian, G. S., Tu, J. T., Groopman, J. D., Gao, Y. T., & Henderson, B. E. (1992). Urinary aflatoxin biomarkers and risk of hepatocellular carcinoma. *Lancet, 339*(8799), 943–946.

Rothman, K. J., Adami, H. O., & Trichopoulos, D. (1998). Should the mission of epidemiology include the eradication of poverty? *Lancet, 352*(9130), 810–813.

Rundle, A., et al. (2007). The urban built environment and obesity in New York City: A multilevel analysis. *American Journal of Health Promotion, 21*(Suppl. 4), 326–334.

Sandel, M. (2005). *Public philosophy*. Cambridge, MA: Harvard University Press.

Sandel, M. (2010). *Justice*. London: Penguin Books.

Sandel, M. (2012). *What money can't buy: The moral limits of markets.* London: Penguin Books.

Sanderson, S. C., et al. (2009). Responses to online GSTM1 genetic test results among smokers related to patients with lung cancer: A pilot study. *Cancer Epidemiology, Biomarkers and Prevention, 18*(7), 1953–1961.

Scoccianti, C., et al. (2011). Methylation patterns in sentinel genes in peripheral blood cells of heavy smokers: Influence of cruciferous vegetables in an intervention study. *Epigenetics, 6*(9), 1114–1119.

Siegel, R., Naishadham, D., & Jemal, A. (2013). Cancer statistics, 2013. *CA: A Cancer Journal for Clinicians, 63*(1), 11–30.

Siraj, A. S., et al. (2014). Altitudinal changes in Malaria incidence in highlands of Ethiopia and Colombia. *Science, 343*(6175), 1154–1158. http://tinyurl.com/qg8lsz7

Soros, G. (2012, September 27). The tragedy of the European Union and how to resolve it. *The New York Review of Books,* 87–93.

Stuckler, D., McKee, M., Ebrahim, S., & Basu, S. (2012). Manufacturing epidemics: The role of global producers in increased consumption of unhealthy commodities including processed foods, alcohol, and tobacco. *PLoS Medicine, 9,* e1001235.

Sullivan, R., Homberg, L., & Purushotham, A. D. (2012). Cancer risk and prevention in a globalised world: Solving the public policy mismatch. *European Journal of Cancer, 48*(13), 2043–2045.

Sunstein, C. R. (2013, March 7). It's for your own good! *New York Review of Books.*

Sunstein, C. R. (2016, January 14). Parking the big money. *New York Review of Books.*

Sylla, B. S., & Wild, C. P. (2012). A million Africans a year dying from cancer by 2030: What can cancer research and control offer to the continent? *International Journal of Cancer, 130*(2), 245–250.

Tang, D., Wang, C., Nie, J., Chen, R., Niu, Q., Kan, H., Chen, B., Perera, F., & Taiyuan, C. D. C. (2014). Health benefits of improving air quality in Taiyuan, China. *Environment International, 73,* 235–242. doi:10.1016/j.envint.2014.07.016.

Vineis, P., Ahsan, H., & Parker, M. (2005). Genetic screening and occupational and environmental exposures. *Occupational and Environmental Medicine, 62*(9), 657–662, 597.

Vineis, P., & Christiani, D. C. (2004). Genetic testing for sale. *Epidemiology, 15*(1), 3–5.

Vineis, P., Kelly-Irving, M., Rappaport, S., & Stringhini, S. (2016). The biological embedding of social differences in ageing trajectories. *Journal of Epidemiology and Community Health, 70*(2), 111–113. doi:10.1136/jech-2015-206089.

Vineis, P., & Khan, A. (2012). Climate change-induced salinity threatens health. *Science, 338*(6110), 1028–1029.

Vineis, P., Stringhini, S., & Porta, M. (2014). The environmental roots of non-communicable diseases (NCDs) and the epigenetic impacts of globalisation. *Environmental Research, 133,* 424–430. http://tinyurl.com/n4ecuhe

Vineis, P., & Wild, C. P. (2014). Global cancer patterns: Causes and prevention. *Lancet, 383*(9916), 549–557.

Vineis, P., & Xun, W. (2009). The emerging epidemic of environmental cancers in developing countries. *Annals of Oncology, 20*(2), 205–212.

Vlachadis, N., et al. (2014). Mortality and the economic crisis in Greece. *Lancet, 383* (9918), 691. http://tinyurl.com/mbmlfnx

von Grebmer, K., Torero, M., Olofinbiyi, T., Fritschel, H., Wiesmann, D., Yohannes, Y., et al. (2011). *Global hunger index. The challenge of hunger: Taming price spikes and excessive food price volatility.* IFPRI, Concern Worldwide, and Welthungerhilfe: Bonn, Washington DC, and Dublin.

WCRF/AICR. http://tinyurl.com/o87dffx

Weiss, M., Middleton, J., & Schrecker, T. (2015). Warning: TTIP could be hazardous to your health. *Journal of Public Health, 37*, 367–369.

West, J. J., Smith, S. J., Silva, R. A., Naik, V., Zhang, Y., Adelman, Z., Fry, M. M., Anenberg, S., Horowitz, L. W., & Lamarque, J. F. (2013). Co-benefits of global greenhouse gas mitigation for future air quality and human health. *Nature Climate Change, 3*(10), 885–889.

Whitehead, M., et al. (Eds.). (2001). *Challenging inequities in health: From ethics to action.* New York: Oxford University Press.

Whitmee, S., Haines, A., Beyrer, C., Boltz, F., Capon, A. G., de Souza Dias, B. F., Ezeh, A., Frumkin, H., Gong, P., Head, P., Horton, R., Mace, G. M., Marten, R., Myers, S. S., Nishtar, S., Osofsky, S. A., Pattanayak, S. K., Pongsiri, M. J., Romanelli, C., Soucat, A., Vega, J., & Yach, D. (2015). Safeguarding human health in the Anthropocene epoch: Report of The Rockefeller Foundation-Lancet Commission on planetary health. *Lancet, 386*, 1973–2028. doi:10.1016/S0140-6736(15)60901-1.

Wild, C. P. (2012). The role of cancer research in noncommunicable disease control. *Journal of the National Cancer Institute, 104*(14), 1051–1058.

Wild, S., Roglic, G., Green, A., Sicree, R., & King, H. (2004). Global prevalence of diabetes. Estimates for the year 2000 and projections for 2030. *Diabetes Care, 27* (5), 1047–1053.

Wilder-Smith, A., & Tambyah, P. A. (2007). The spread of poliomyelitis via international travel. *Tropical Medicine and International Health, 12*(10), 1137–1138.

Woodcock, J., Edwards, P., Tonne, C., Armstrong, B. G., Ashiru, O., Banister, D., Beevers, S., Chalabi, Z., Chowdhury, Z., Cohen, A., Franco, O. H., Haines, A., Hickman, R., Lindsay, G., Mittal, I., Mohan, D., Tiwari, G., Woodward, A., & Roberts, I. (2009). Public health benefits of strategies to reduce greenhouse-gas emissions: Urban land transport. *Lancet, 374*(9705), 1930–1943. doi:10.1016/S0140-6736(09)61714-1.

Zhang, F. F., et al. (2011). Physical activity and global genomic DNA methylation in a cancer-free population. *Epigenetics, 6*(3), 293–299.

Index

© Springer International Publishing AG 2017
P. Vineis, *Health Without Borders*, DOI 10.1007/978-3-319-52446-7

105